Network Analysis
for Technology

CURTIS D. JOHNSON

University of Houston

NETWORK ANALYSIS FOR TECHNOLOGY

Macmillan Publishing Company

New York

Collier Macmillan Publishers

London

Macmillan Publishing Company
866 Third Avenue, New York, New York, 10022

Collier Macmillan Canada, Inc.

Library of Congress Cataloging in Publication Data

Johnson, Curtis D.,
  Network analysis for technology.
  Includes index.
  1. Network analysis (Planning) I. Title.
T57.85.J57 1984   621.319′2   83-847
ISBN 0-02-361050-6

Printing: 1 2 3 4 5 6 7 8   Year: 4 5 6 7 8 9 0 1

ISBN   0-02-361050-6

PREFACE

This text was written for a single semester advanced network analysis course for Technology or Applied Engineering majors. The basic objective of the text is to impart to the student sufficient understanding of advanced network concepts and math to enable him or her to solve any linear electrical/electronic network. This is followed by an introduction to Fourier series, integral transforms, and discrete transforms, to provide the student with the basic ability to interpret signals resulting from analysis. These task-oriented objectives, when achieved, also leave the student with a deeper knowledge and intuitive grasp of network behavior. Such comprehension provides a subtle but essential understanding to aid the students in following courses and careers in electrical/electronic fields.

The treatment of network analysis presented by this text maintains the practical goal of Technology by presenting the network and supporting math topics as result-generating objectives. As much supporting theory and derivation is presented as necessary to maintain continuity of the material.

It is assumed that the student has had courses in elementary dc and ac analysis, algebra, trigonometry, and differential calculus. Integral calculus should either be a corequisite or prerequisite. Classical differential equations is not required. Sufficient study of differential equations is presented in the text to enable the student to see how the equations are generated and to solve simple network problems. The fundamental approach of the text for solving network problems is to use Laplace transforms in combination with MESH and NODAL matrix constructions. This is the most error free and efficient method of setting up and solving general linear networks. Since many Technology programs do not include matrix algebra in algebra courses, this subject, along with the evaluation of higher order determinants, is presented in the text.

Use of the computer is emphasized throughout the text as a tool to aid in solving network problems. This is not a text in numerical methods and so the algorithms are not presented. Rather it is assumed that a computer with a math package of programs is available. An instructor manual with programs in BASIC as necessary for this text is available.

The structure of the text follows a design of building the tools of the student to the point where, in Chapter 5, everything can be pulled together, culminating in the ability to set-up the equations and solve any linear, time dependent network. Chapter 1 is a review of basic network definitions and concents. It should be material which students have seen in other courses. The purpose of the chapter is to pull many concepts together and provide consistent definitions. It may be possible in some programs to assign this as reading material with the problems for review.

Chapter 2 builds the basic skills of the student in MESH and NODAL matrix approaches for solving linear networks with only resistive components. Controlled sources are presented, as well as applications of network analysis for finding Thévenin and Norton equivalent circuits.

Chapter 3 shows the student the effects of introducing capacitors, inductors, and transformers into the networks. The concepts of initial conditions are covered and the emergence of integro-differential equations from the networks is explored. Classical first and second order differential equations are developed as a technique for solving simple networks with capacitors and/or inductors and transformers.

Chapter 4 develops the technique of Laplace transforms as a method of solving the integro-differential equations which result from networks with capacitors and/or inductors and transformers.

Chapter 5 is the culmination of the text for network analysis. In this chapter the methods of generalized MESH and NODAL analysis are presented by which the network is transformed into s-space and then matrix equations written, almost by inspection. After completing this chapter the student will have the skills to solve any linear network with resistors, capacitors, inductors, transformers, controlled sources, and time-dependent real sources.

Chapter 6 presents the basic elements of Fourier series to aid the student in interpretation of network analysis results and in some cases the analysis itself. The concepts of Fourier integral transforms and discrete transforms are presented in introductory form so the student will be prepared for more detailed treatment in other courses.

I wish to thank my wife, Helene Blake, and son, Greg Johnson, for their patience with me during the writing of this text. I know a lot of things got left undone during this period. There is this nagging suspicion that they are still waiting for me. That's ok.

C.D.J.

CONTENTS

CHAPTER 1

Introduction to Network Analysis

Objectives

The primary objectives of this chapter are to review the basic network concepts and to present formal definitions of network elements and laws. After studying this chapter and doing the problems at the end of the chapter, you will be able to:

1. Define an electrical network in terms of the elements and components that make up the network.
2. Explain the electrical characteristics of the common network elements: resistors, capacitors, inductors, transformers, current sources, and voltage sources.
3. Explain the four types of controlled sources and the principle of their application in network analysis.
4. Define the two basic network laws, KVL and KCL.
5. Define and diagram the common steady-state signals that occur in electrical networks.
6. Define and diagram the common transient signals that occur in electrical networks.

NETWORK ANALYSIS

The first thing that must be done to embark on a study of network analysis is to *define* exactly what we mean by a *network* and by the *analysis* of that network.

1-1.1 Network

A *network* is simply an electrical interconnection of objects. The objects are called the *elements* of the network, so a network is often called an *assemblage of elements*. There are many modifiers to give more information about what *kind* of network is under consideration. For example, the network may be active, passive, linear, or nonlinear. Definition of these terms and others is part of the objective of this chapter.

1-1.2 Network Analysis

Network analysis is the systematic application of specified laws and principles to an assemblage of electrical elements, a network, which allows a complete prediction of the electrical behavior of that assemblage. This means that the result of the analysis is complete knowledge, for all time covered by the analysis, of the network electrical properties.

To *predict* the electrical behavior of a network means that there is a set of variables which describe that behavior *and* that the values these variables will take in the future can be determined. Indeed, network analysis *solves* for the values of these variables, usually as functions of time. These time functions specify the value of the variables at any future time covered by the solution.

It is important to realize that approximations are often made which place limitations on the validity of the solutions. These limitations are both in time and in the range of network variable values. One of the most common approximations is the replacement of nonlinear circuit elements by linear approximations.

1-1.3 Network Variables

There are only two network variables necessary to describe the behavior of an assemblage of electrical elements. These variables are the *voltages* across elements and the *currents* through elements. Actually, as you will learn later, complete knowledge of only one set of these variables is sufficient to describe the electrical behavior of a network. When a network is *solved*, the values of these variables have been determined, either numerically or by equations of time. Figure 1–1 presents a network where the elements have been represented by rectangles. The network variables are shown to be the set of currents, i_1, i_2, . . ., i_8, which flow through each element and the set of voltages, v_1, v_2, . . ., v_5, at each point of connection of the elements with respect to a defined reference.

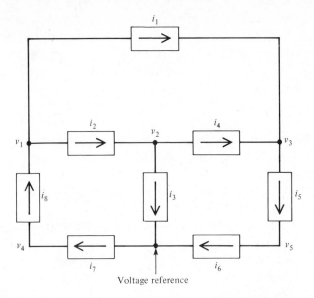

FIGURE 1–1. A network is an assembly of elements with currents and voltages as the network variables.

Voltage. Voltage is a measure of potential energy difference between two points. That is why a voltage specification must always be referred to some other point, as in Fig. 1–1, where a reference was defined with respect to which the voltage of various points in the network is referred. The *voltage drop* across some element in the network is simply the voltage on one side of the element with respect to the other. In Fig. 1–1 the voltage drop across the element with current i_2 is thus given by the difference, $v_1 - v_2$. We use the lowercase letter v to indicate a general voltage, which may be a function of time, $v(t)$. The proper unit of voltage is the *volt*, whose symbol is V. Voltage has a *polarity* indicating which parts of the circuit are at higher potential energy than other parts. In Fig. 1–1, if $v_1 - v_2 = +10$ V, it would mean that the point at v_1 is at a higher potential than the point at v_2. Conversely, if the difference were -10 V, the point at v_2 would be at the higher potential.

The concept of a *ground* often appears in networks, although we do not use it in this text. The ground is simply that point of the network which has been chosen to be the reference. If some other point in the network has a negative voltage with respect to the ground, or reference, it simply means that the point chosen for a reference is at a higher potential. Any point in a network can be chosen for the reference.

Current. Current is a measure of the flow of charges through an element of the network. The unit of current is the *ampere*, whose symbol is A. One ampere is defined as one coulomb of charge per second, reflecting the fact that it is a measure of charge flow.

There are two ways of thinking of current. In one instance, called *electron flow*, the current is described by the actual charge carrier which most often occurs in electricity, the electron. According to this definition, current would flow from the points in the network of negative potential (excess electrons) to the points of positive potential. This is a difficult concept to visualize if we think of positive as *higher* potential. To be more consistent with energy concepts, it has become common to define charge flow in terms of *conventional current* flow. By this definition, which is used in this text, current is simply *defined* to flow from points of higher potential (voltage) to points of lower potential. The specific nature of the charge carrier is ignored. In Fig. 1–1, if the current $i_4 = +2$ A and flows in the direction shown, it would mean that v_2 was larger (more positive) than v_3.

There are many textbooks and technical disciplines which use electron flow to describe electrical behavior. In these cases a comparison can be made with conventional current flow descriptions by changing the sign of the currents.

1-1.4 Network Energy

The electrical behavior of a network is actually a specification of the energy and energy flow within the elements of the network. When a network has been solved by finding the voltages and currents associated with the elements, it is possible to determine the energy stored or rate of energy gain or loss of each element. Often this is the objective of solving the network.

Power. The power, which is the rate of energy absorbed or generated by an element in a network, can be found as a product of the current through and the voltage drop across the element:

$$p(t) = i(t)v(t) \qquad\qquad (1\text{-}1)$$

where

$$p(t) = \text{power, watts (W; joules/second)}$$

$$i(t) = \text{current through the element, amperes}$$

$$v(t) = \text{voltage across the element, volts}$$

Note that the power of Eq. (1-1) is *instantaneous* since it gives the power associated with the element at every instant of time. The value given by Eq. (1-1) is the energy *delivered to the network* by the element if the current flows *from* the side of the element at greater potential. It is the energy *delivered to the element* or *dissipated* from the network if the current flows *into* the side of the element at greater potential.

Energy. The energy associated with an element of the network is usually dynamic, that is, changing in time. For this reason it is often more appropriate to speak of the energy flow, or power, defined above. In some cases, however, it is of value to determine the energy. Since we know that the power is the derivative of the energy, $p(t) = dw/dt$, the energy can be found by integration of the power associated with an element over the time period of interest.

$$w(t) = \int_{t_0}^{t} p(\tau) \, d\tau + w(t_0) \qquad (1\text{-}2)$$

where

$$\tau = \text{variable of integration}$$

$$w(t) = \text{energy at time } t, \text{ joules}$$

$$w(t_0) = \text{energy at time } t_0, \text{ joules}$$

1-2

PASSIVE NETWORK COMPONENTS

Network analysis is a very systematic process and requires a common understanding, or agreement, on the meaning of the symbols, terms, and concepts employed. In Fig. 1–1, the elements that make up the network are represented as rectangles. These elements may consist of many different objects. In this and the next two sections the various types of objects are reviewed and defined.

Elements can be broken down into *components* and *sources*. Components are elements that cannot inject new energy into the network. Sources are elements that inject new energy into the network. Components can be further divided into *active* and *passive* types. Active components can act as *channels* through which sources can inject energy into circuits. Active components can change the form of energy in a network. Passive components, such as resistors, capacitors, inducators, and transformers, can only store or dissipate energy. A resistor dissipates energy from a network, whereas capacitors, inductors, and transformers can store and release energy, but none of these objects can channel energy from a source or change the form of the energy. Active components include diodes, transistors, operational amplifiers, and a host of other special devices. To work effectively in network analysis it is essential to have a very clear idea of the electrical characteristics of the elements. The following sections give a brief review and formal definitions of passive network components.

1-2.1 Resistor

Certain elements in networks obey a property that the ratio of the voltage drop across the element and the current through the element is a constant. This means if the voltage is doubled, the current would double so that the ratio remains constant. The constant ratio is defined to be the *resistance* and the device is called a *resistor*. The unit of resistance is the *ohm*, which is defined as a ratio of one volt to one ampere and has the symbol Ω. Analytically, we describe the resistor as any element that obeys Ohm's law that the ratio of voltage to current is a constant, or

$$v(t) = i(t)R \tag{1-3}$$

where

$$v(t) = \text{voltage across the resistor}$$

$$i(t) = \text{current through the resistor}$$

$$R = \text{resistance, ohms}$$

Note that the resistor is an energy-dissipative device and therefore the current will always be directed into the side of the resistor with higher potential (voltage). This is shown in Fig. 1–2, which gives the schematic symbol for the resistor. This figure shows that the voltage drop is positive on the side of the resistor which the current enters. Equation (1-3) shows that the resistor voltage drop and the current are in phase and you should also remember that the ideal resistor used in network analysis has zero capacity and zero inductance.

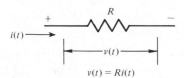

FIGURE 1–2.   Ideal network resistor and variable relations.

Using Eqs. (1-1) and (1-3), it is easy to show that the power dissipated by the resistor can be expressed as

$$p(t) = i^2(t)R = \frac{v^2(t)}{R} \tag{1-4}$$

This power is dissipated as heat.

EXAMPLE 1-1

A resistor of 20 Ω has a direct current of 4 A. What is the voltage across the resistor and the power dissipated?

SOLUTION

From Eq. (1-3) we find that $v = iR = (4)(20) = 80$ V. Then using Eq. (1-4) we get $p = i^2R = (4)^2(20)$ or $p = 320$ W. ∎

EXAMPLE 1-2

A resistor is found to have a voltage drop given by $v(t) = 100e^{-t/4}$ and a current of $i(t) = 2e^{-t/4}$. Find the value of the resistance and the instantaneous power dissipated by the resistor.

SOLUTION

The resistance can be found by noting, from Ohm's law, that the ratio of voltage to current is a constant, equal to the resistance. Then we find $R = v/i = 100e^{-t/4}/2e^{-t/4} = 50$ Ω. The power is found from Eq. (1-1):

$$p = iv = 200e^{-2t/4} = 200e^{-t/2} \text{ W}$$ ∎

Note that typical values of resistances in networks may vary from a few ohms to many millions of ohms. To make certain that we do not miss the point of network analysis technique because of numerical complexity, in this text we always use values of resistance of less than 100 Ω.

Wire. In setting up a network to be solved by network analysis, the various elements are connected by lines representing the wires that would connect a real circuit. In our analysis we will assume that these wires have *zero resistance*. There are called *ideal* wires. If the problem to be solved is derived from some real-world circuit, it may be necessary to include the finite resistance of the wires actually used. In this case a *model* will be constructed that has all the electrical characteristics of the real wire. Figure 1–3 shows that a real wire with resistance R_w is modeled by ideal wires and a resistor of value R_w.

1-2.2 Capacitor

Analytically, a capacitor is a component for which the ratio of stored electric charge $q(t)$ to voltage drop $v(t)$ is constant. This means that if the amount of

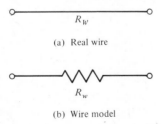

(a) Real wire

(b) Wire model

FIGURE 1–3. A real wire can be modeled by an ideal wire and a resistor.

charge is doubled, the voltage will double in such a way that the ratio remains constant. This ratio is called the *capacity* of the device. The unit of capacity is called the *farad* (unit symbol F), which is equal to one coulomb per volt. In equation form the capacity is

$$C = \frac{q(t)}{v(t)} = \text{constant} \tag{1-5}$$

where C is the capacity in farads. This relationship is not very useful in network analysis since we do not deal with charges themselves but rather the rate of flow of charges (i.e., currents). But, of course, current is just the derivative of charge, $i(t) = dq/dt$. If Eq. (1-5) is solved for the charge and the resulting equation is differentiated, the result is a relationship between voltage and current:

$$i(t) = C \frac{dv}{dt} \tag{1-6}$$

This equation shows that the current through a capacitor at any time is given by the rate at which the voltage across the capacitor is changing, times the value of the capacity.

The schematic symbol of a capacitor is shown in Fig. 1–4, together with the relationship between current direction and voltage drop polarity. The ideal capacitor has an *infinite* resistance and *zero* inductance.

A resistor was shown to dissipate energy from a network. In the case of a capacitor, energy is *stored* and no energy is dissipated. This ability of the capacitor to store energy, which can subsequently be reclaimed by the network, is the basis for the usefulness of the element in networks. Using Eqs. (1-1) and (1-6), the power associated with a capacitor can be written as $p(t) = Cv(t)\,dv/dt$. If this expression is used in Eq. (1-2), the energy stored in a capacitor can be found as follows:

$$w(t) = C \int_{t_0}^{t} v \frac{dv}{dt}\, dt + w(t_0)$$

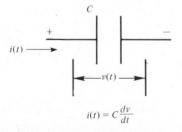

FIGURE 1–4. Ideal network capacitor and variable relations.

For convenience let us set $w(t_0) = 0$, which simply means that the capacitor was initially uncharged. The integral can be performed using integration by parts. $\int x\, dy = xy - \int y\, dx$. If $x = v$ and $dy = (dv/dt)\, dt$, you can see that $dx = dv$ and $y = v$. The integration now becomes

$$w(t) = C\left(v^2 - \int_{v(t_0)}^{v(t)} v\, dv\right)$$

or

$$w(t) = C(v^2 - \tfrac{1}{2}v^2)$$

where, since $w(t_0) = 0$, then $v(t_0) = 0$ also. This gives a final result of

$$w(t) = \tfrac{1}{2}Cv^2(t) \tag{1-7}$$

$$\text{where } w(t) = \text{energy stored at time } t$$

$$v(t) = \text{voltage on the capacitor at time } t$$

Equation (1-7) shows that the amount of energy stored at time t depends only on the voltage on the capacitor at that time.

EXAMPLE 1-3

A 2-F capacitor has a slowly increasing voltage across its terminals. The voltage rises at a rate of 0.02 V per second (i.e., $v = 0.02t$). Find the current through the capacitor and the energy stored after 3 seconds(s).

SOLUTION

The current can be found from Eq. (1-6):

$$i = C\frac{dv}{dt} = (2)\frac{d}{dt}(0.02t) = (2)(0.02) = 0.04 \text{ A}$$

Thus the current is a steady (direct current) 0.04 A. The energy stored after 3 s is found from Eq. (1-7). The voltage is $v = 0.02t = (0.02)(3) = 0.06$ V. Then, from Eq. (1-7), $w = \tfrac{1}{2}Cv^2 = (2)(0.06)^2/2$ or $w = 0.0036$ joule (J). ∎

Typical values of capacity vary from a few picofarads (10^{-12} F) to a few millifarads (10^{-3} F) in real-world circuits. In an effort to simplify the numerical problems and concentrate on the network analysis concepts, capacity values of 1 to 10 F will be used in this text.

1-2.3 Inductor

The inductor, like the capacitor, is a network element which can *store* electrical energy. The capacitor stores energy in the form of an electric field between charges deposited in the device. The inductor stores the energy in the form of a magnetic field generated by a current flowing through a coil of wire. An analytical definition of the inductor cannot be deduced from first principles but rather comes from an *observation* of the relationship between current and voltage associated with a coil of wire. In particular, it is found that the ratio of voltage across an inductor (a coil of wire) to the rate of change of current through the element is a constant. We call this constant the *inductance*. The unit of induct- ance is the *henry*, which is denoted by the symbol H. In equation form the voltage across an inductor is given by

$$v(t) = L \frac{di}{dt} \tag{1-8}$$

where L is the inductance in henrys. Figure 1-5a shows the schematic symbol for the inductor and the relationship between current direction and voltage drop polarity associated with this element.

An ideal inductor has *zero* resistance from terminal to terminal and *zero* capacity. In actual practice the inductor is made from wire, which has some finite resistance. Thus we often must use an *analytical model* of an inductor when studying real circuits. Such a model, as shown in Figure. 1–5b, will use an ideal inductor in series with a resistor, representing the wire resistance. Of course, the junction between inductor and resistor does not exist since the re- sistance is distributed throughout the inductor. Nevertheless, this model shows all the circuit behavior of a real inductor with finite resistance and so can be used in network analysis to replace the inductor.

The energy stored in an inductor can be reclaimed by the network, like that of the capacitor, and it is this storage and reclamation of energy that makes the

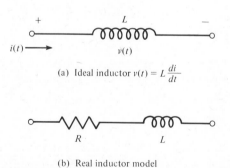

(a) Ideal inductor $v(t) = L \dfrac{di}{dt}$

(b) Real inductor model

FIGURE 1–5. Ideal inductor and variable relations. A real inductor is modeled by an ideal inductor and a resistor.

inductor useful in networks. But unlike the capacitor, the energy storage in an inductor is *dynamic* in the sense that it is stored only when current is flowing. The capacitor may be "charged" and then placed on a shelf and the energy stored will remain. You cannot do this with an inductor. In general, a *real* inductor will also dissipate some energy because of the resistance of the wires of which it is composed. We can find an expression for the stored energy from the relation

$$p(t) = i(t)v(t) = Li(t)\frac{di}{dt}$$

and Eq. (1-2),

$$w(t) = \tfrac{1}{2}Li^2(t) \tag{1-9}$$

Note that the instantaneous energy stored, $w(t)$, is a function of the current flowing in the inductor.

EXAMPLE 1-4
An inductor of 4 H has a steady current of 2.5 A. What is the voltage across the inductor and the energy stored?

SOLUTION
We get the voltage from Eq. (1-8), $v = L(di/dt)$ or $v = (4)[d(2.5)/dt] = 0$ V! Of course, the voltage is zero since the current is not changing. The energy stored is found from Eq. (1-9), $w = \tfrac{1}{2}(4)(2.5)^2$ or $w = 12.5$ J. ∎

EXAMPLE 1-5
Suppose that the inductor of Example 1-4 has a finite resistance of 0.02 Ω. Find the voltage, stored energy, and power dissipated.

SOLUTION
In this case we can first see that the stored energy will be the same since the "ideal" inductor in the model, Fig. 1–5b, still has a current of 2.5 A. The voltage will not be zero now since there will be some voltage drop across the resistance. In particular, from Ohm's law, $v = (2.5)(0.02) = 0.05$ V. The dissipation is then $p = (2.5)^2(0.02)$ or $p = 0.125$ W. This means that it "costs" 0.125 J per second to store 12.5 J in the inductor. ∎

1-2.4 Transformer
The transformer is a passive circuit element which couples two parts of a network by magnetic fields. Figure 1–6 shows a transformer connected between networks A and B. *Coupling* means that variations in i_1 and v_1 will change the values of i_2 and v_2, and vice versa. These changes are caused because the magnetic flux

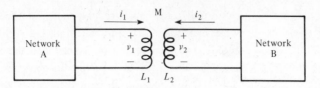

FIGURE 1–6. Ideal transformer and relations between network variables. The transformer couples networks A and B.

from each coil of the transformer passes through the other coil. To describe the action of the transformer analytically, it will be necessary to use a parameter that describes the degree to which this coupling of magnetic flux occurs. This will then provide equations that relate the two sets of currents and voltages. Such equations can be derived from simple considerations of how the coupling occurs.

Figure 1–6 shows that there are three characteristics of the transformer. Each coil has an inductance, called the *self-inductances*, L_1 and L_2, which are just as described in Section 1-2.3. In addition, there is a *mutual inductance*, M, which is a measure of the *strength* or *degree* of the coupling. Zero mutual inductance would be *no* coupling, resulting in two independent inductors. The maximum value of M turns out to be given by $(L_1L_2)^{1/2}$. Then the mutual inductance can be described by an equation, $M = k(L_1L_2)$, where k is the coupling constant. For different transformers k will have values ranging from 0 to 1, depending on the degree of coupling.

Suppose that the mutual inductance is zero, $M = 0$ (this could be provided by moving the coils a great distance apart). Then we simply have two inductors and there is no coupling; hence, from Eq. (1-8),

$$v_1(t) = L_1 \frac{di_1}{dt}$$

and

$$v_2(t) = L_2 \frac{di_2}{dt}$$

Now suppose that the coils are moved closer together, so that some coupling occurs (i.e., $M > 0$). Then some of the magnetic flux from coil 1 will pass through coil 2, and vice versa. This will affect the coil voltages, just like an inductor via Eq. (1-8), except that the coupling will be described by the mutual inductance M instead of the self-inductances. The contribution will simply add on to the effect of the self-inductances. The voltage equations will then become

$$v_1(t) = L_1 \frac{di_1}{dt} + M \frac{di_2}{dt} \qquad (1\text{-}10)$$

$$v_2(t) = L_2 \frac{di_2}{dt} + M \frac{di_1}{dt} \qquad (1\text{-}11)$$

Equations (1-10) and (1-11) can be used to represent transformers in networks to be solved. However, another important feature to remember is that the *effect* of mutual inductance can be *positive* or *negative*. The reason for this is that the two coils may not be *wound* in the same sense, clockwise or counterclockwise, on whatever form is used to construct the transformer. The result will be that a positive rate of change of, say, i_2 in Eq. (1-10), may cause v_1 to increase or decrease. Positive and negative connections for mutual inductance are indicated in a schematic diagram by dots placed near the coils. In the transformer of Fig. 1–7a the dots are on the same side of the coils, which means that the effect of mutual inductance is positive. In Fig. 1–7b the dots are on opposite sides of the coils, showing that the mutual inductance produces a negative effect on voltage.

Equations (1-10) and (1-11) can be used, with appropriate signs to indicate a positive or negative connection of the coils, to solve network problems involving transformers. In some cases it is possible, and easier, to use a model of the transformer to simplify the equations. There are two models of transformers which are often used in network analysis, the ideal model and the T model.

Ideal Model.   In this model the self- and mutual inductances are assumed to be infinite and the coupling is assumed to be perfect. This means that all of the flux from one coil goes through the other coil. In this case the relationships that relate current and voltage depend *only* on the number of turns in the two

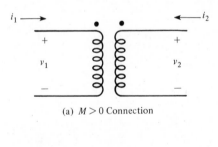

(a) $M > 0$ Connection

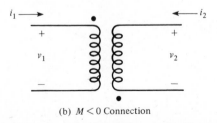

(b) $M < 0$ Connection

FIGURE 1–7.   A transformer can have either positive or negative mutual inductance, which is indicated schematically by dots as shown.

coils and the polarity of the coupling. For positive connections of mutual inductance, we have

$$\frac{v_2(t)}{v_1(t)} = \frac{n_2}{n_1} \quad \text{and} \quad \frac{i_2(t)}{i_1(t)} = -\frac{n_1}{n_2} \tag{1-12}$$

where

$$n_1 = \text{number of turns in coil 1}$$

$$n_2 = \text{number of turns in coil 2}$$

If the mutual inductance effect is negative, the signs of the two relations in Eq. (1-12) are reversed.

Certain types of transformers, closely wound on iron cores, fit this ideal model rather well.

T Model. Another model used in network analysis has equations exactly like Eqs. (1-10) and (1-11) but is modeled from inductors in a "tee" connection, as shown in Fig. 1–8. This model is used when the two bottom leads of the transformer have been connected. The effective values of the "inductors" in this model are found from the transformer characteristics as follows:

For *positive* connected mutual inductance we have

$$L_A = L_1 - M$$

$$L_B = L_2 - M \tag{1-13}$$

$$L_C = M$$

For *negative* connected mutual inductance we have

$$L_A = L_1 + |M|$$

$$L_B = L_2 + |M| \tag{1-14}$$

$$L_C = -|M|$$

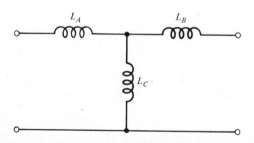

FIGURE 1–8. T model of a transformer.

If the resistance of the wires is important to the analysis, the T model would include resistors in series with L_A and L_B to represent these resistances.

EXAMPLE 1-6

A transformer has self-inductances of 2 H and 4.5 H and a positive mutual inductance of 1.5 H. Find the T model of the transformer. Use Eqs. (1-10) and (1-11) to find the voltages if the currents are given as $i_1 = 5t$ and $i_2 = -2t$.

SOLUTION

The T model is found using the relations of Eq. (1-13). Referring to Fig. 1–8, the values of inductances are: $L_A = 2 - 1.5 = 0.5$ H, $L_B = 4.5 - 1.5 = 3$ H, and $L_C = 1.5$ H. The voltages for the given currents are

$$v_1(t) = 2 \frac{d(5t)}{dt} + 1.5 \frac{d(-2t)}{dt}$$

$$v_1(t) = 2(5) + 1.5(-2) = 7 \text{ V}$$

and

$$v_2(t) = 4.5 \frac{d(-2t)}{dt} + 1.5 \frac{d(5t)}{dt}$$

$$v_2(t) = 4.5(-2) + 1.5(5) = -1.5 \text{ V.} \qquad \blacksquare$$

1-3

NETWORK SOURCES

Network sources represent the elements that provide the energy which allows a network to function. These elements may be constant or vary in time. Examples include the common battery, direct-current (dc) power supplies, and the various types of alternating-current (ac) signal generators. In network analysis, sources are separated into two types, the voltage source and the current source. In this section we review the basic characteristics of each type and discuss the process of transforming one type of source into the other.

1-3.1 Ideal Voltage Source

A *voltage source* is a source of energy characterized by the property that the voltage across its terminals is maintained at some specified value. An actual voltage source, such as a dc battery, deviates somewhat from this definition because of the existence of an *internal resistance* in the source. This means that under various loads the voltage at the terminals of the source will not be the expected value. This situation would make analysis of a network very difficult.

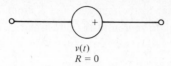

$v(t)$
$R = 0$

FIGURE 1–9.   Ideal network voltage source.

For this reason we define the *ideal voltage source*, shown in Fig. 1–9. This element has *zero* internal resistance. The ideal source will maintain its terminals at the indicated voltage no matter what load is connected. *Real sources* can be modeled using the ideal source. For example, as illustrated in Fig. 1–10, a battery with some specified internal resistance would be modeled by an ideal source in series with a resistor equal to the internal resistance, and a signal generator with a stated 600-Ω resistance is modeled by a voltage source in series with a 600-Ω resistor.

EXAMPLE 1-7

A 1.5-V dry-cell battery is found to have a 5-Ω internal resistance. Using an equivalent model, determine the current drawn by a 10-Ω load and the cell terminal voltage under this load.

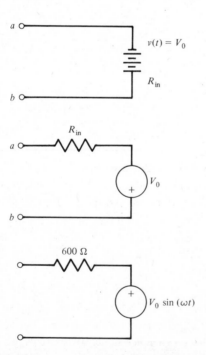

FIGURE 1–10.   A real battery is modeled by an ideal voltage source and a resistor. A signal generator is modeled in the same way.

INTRODUCTION TO NETWORK ANALYSIS

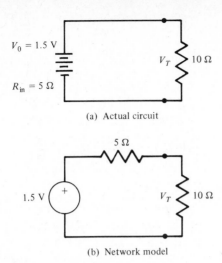

(a) Actual circuit

(b) Network model

FIGURE 1–11. Networks for Example 1-7.

SOLUTION

Figure 1–11 shows the problem as a battery with a 10-Ω load and a battery model with a voltage source in series with an internal resistance. The model shows more clearly the features of the problem. The current is simple to find from Ohm's law: $I = V/R = 1.5/(5 + 10) = 100$ mA. The cell terminal voltage is found by subtracting the voltage dropped across the internal resistance: $V_T = 1.5 - (0.1)(5) = 1.0$ V. So the cell would measure only 1 V when loaded by 10 Ω. ■

Series and Parallel. It is possible to connect ideal voltage sources in series and the total voltage will then be the algebraic sum of the sources. In Fig. 1–12a the voltage of the series is $+6$ V, while in Fig. 1–12b the series combination is only $+2$ V. It is *not* possible to connect ideal voltage sources

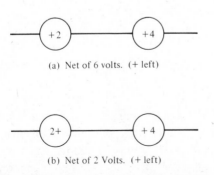

(a) Net of 6 volts. (+ left)

(b) Net of 2 Volts. (+ left)

FIGURE 1–12. Ideal voltage sources can be placed in series in two ways.

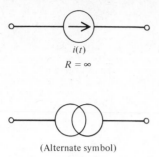

FIGURE 1–13. Ideal network current source.

in parallel, unless they have exactly the same voltage, since there can be only one voltage value between two points in a network.

1-3.2 Ideal Current Source

A current source is not as common as a voltage source and yet it is very often used in network analysis. In network analysis we use an *ideal current source*, a device that will deliver a specified number of amperes into an attached load, regardless of the nature of the load. The voltage across the terminals of a current source can be any value appropriate to support delivery of the specified current. The schematic symbol of a current source is shown in Fig. 1–13. Note that the internal resistance of this ideal source is infinite (i.e., an open circuit). Actual current sources used in circuits have finite internal resistance. When it is necessary to include this internal resistance, the real source is modeled by an ideal source shunted by a resistor of the specified internal resistance, as shown in Fig. 1–14.

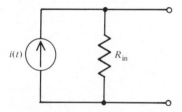

FIGURE 1–14. A real current source is modeled by an ideal source and a resistor.

Series and Parallel.   It is possible to connect ideal current sources in parallel as shown in Fig. 1–15a. In this case the net current delivered is the algebraic sum of the sources. In Fig. 1–15a the net current is then 8 A, while in Fig. 1–15b the net current is only 2 A. It is not possible to connect ideal current sources in series, unless they have exactly the same value, because all

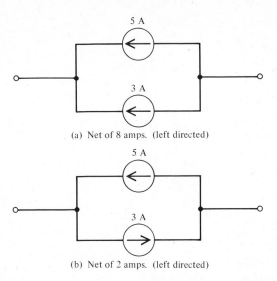

(a) Net of 8 amps. (left directed)

(b) Net of 2 amps. (left directed)

FIGURE 1–15.   Ideal current sources can be placed in parallel in two ways.

the current delivered by one in the series must be taken and delivered by the next.

1-3.3   Source Transformations

In many cases of network analysis it is desirable to use either current sources or voltage sources, exclusively, when finding a solution for the network variables. It is possible to do this because it is possible to *transform* voltage sources to current sources, and vice versa. The transformation can be made even when the source is time dependent. Such a transformation is possible as long as the two representations of an energy source are *analytically equivalent;* that is, they interact in a circuit to which they are connected in exactly the same fashion. In essence you cannot tell them apart by their electrical behavior.

To see how the transformationn is defined, consider the ideal voltage source $v(t)$ in series with a resistor R in Fig. 1–16a. What are the electrical characteristics of this combination, as measured between terminals a and b? Well, a measurement of *resistance* from a to b would yield R, since the source has zero internal resistance by definition. A measurement of *open-circuit voltage* from a to b would yield $v(t)$, since with no current there is no voltage drop across R. Finally, a measurement of *current* through a *short circuit* from a to b would yield $i(t) = v(t)/R$, by Ohm's law.

Now, consider the current source shunted by resistance R in Fig. 1–16b. We have chosen to make the value of the current source $i(t) = v(t)/R$. Notice that the three measurements made above will result in exactly the same values! Resistance will give R since the current source is an open circuit. Open-circuit voltage will give $v(t)$ since all the current will go through R and thus there will be a voltage drop across the resistor of $v(t) = [v(t)/R]R = v(t)$. Of course,

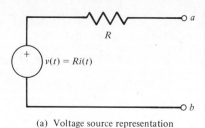

(a) Voltage source representation

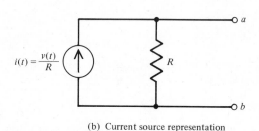

(b) Current source representation

FIGURE 1–16. **A real source can be modeled by either an ideal voltage source or an ideal current source.**

the short-circuit current will be $i(t) = v(t)/R$, since all current will go through the short and none through the resistance R. So it is clear that the two representations of Fig. 1–16 are analytically equivalent since there is no measurement that could distinguish which of the two circuits was being used. The transformation process can be generalized into two rules:

Voltage Source Transformation Rule. *A voltage source in series with a resistance can be replaced by a current source with magnitude equal to the voltage divided by the resistance, shunted by the same resistance. The current source direction will be to the same terminal as the voltage source positive terminal.*

Current Source Transformation Rule. *A current source shunted by a resistance can be replaced by a voltage source with magnitude equal to the product of the current and the resistance, in series with the resistance. The positive of the voltage source will be on the same terminal as that pointed to by the current source.*

EXAMPLE 1-8

Given an ideal voltage source of 4 V in series with a 2-Ω resistor as shown in Fig. 1–17a, transform the voltage source to a current source.

SOLUTION

This will be a current source, $I = 4/2 = 2$ A. The source will be shunted by a 2-Ω resistor, as shown in Fig. 1–17b. ∎

INTRODUCTION TO NETWORK ANALYSIS

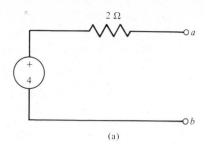

(a)

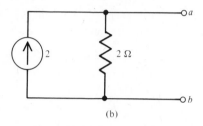

(b)

FIGURE 1–17.   Networks for Example 1-8.

EXAMPLE 1-9

Given a current source of 3 cos (4t) A shunted by a resistor of 6-Ω, express the current source as a voltage source.

SOLUTION

This will be a voltage source, $v(t) = [3 \cos (4t)](6)$ or $v(t) = 18 \cos (4t)$ in series with the 6-Ω resistor, as shown in Fig. 1–18.   ∎

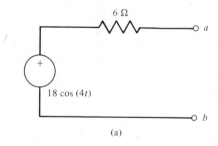

(a)

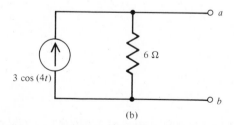

(b)

FIGURE 1–18.   Networks for Example 1-9.

Compound Tranformations. In some cases it may be desirable to transform a voltage source to a current source when no single resistor is in series with the source or to transform a current source to a voltage source when no single resistor is shunting the source. It is possible to perform the transformation in those cases where combinations of resistors are available by finding an analytically equivalent form of the network.

Figure 1–19a shows a 10-V source connected to two resistors, one of 10 Ω and one of 5 Ω. The two resistors are connected in some fashion to the rest of the network and are *not* in parallel. In this case a transformation can be made by writing the voltage source as two sources in parallel and then separating them so that each is connected to a resistor as shown in Fig. 1-19b. This is equivalent since all the 10-V source did was to maintain the left side of the resistors at

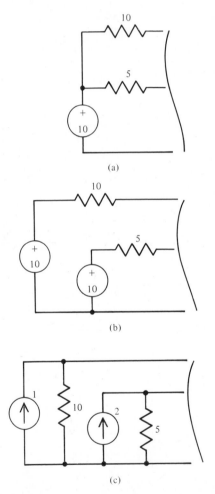

FIGURE 1–19. Example of compound transformation of a voltage source to current sources.

INTRODUCTION TO NETWORK ANALYSIS

10V, and this is true of both representations. Now conversion to current sources is simple, as shown in Fig. 1–19c.

Figure 1–20a shows a 2-A current source shunted by two resistors, of 4-Ω and 3-Ω. The resistors are connected to the rest of the network and are *not* in series. In this case the 2-A source is written as two 2-A sources in series, as shown in Fig. 1–20b. Now point *a* and point *b* of the network can be connected together, since *no current can flow through the connecting wire!* The two 2-A sources in series do not allow any current to go through the wire from *a* to *b*. Now the network has the form of Fig. 1–20c and transformation to voltage sources can be easily made, as shown in Fig. 1–20d.

You should remember that even though analytically equivalent, a transformed circuit is not actually the same as the original. Any calculation of voltage, current, or power of the transformed representation itself will not give physically meaningful results with respect to the actual circuit. It is only the external response of the transformed circuit which is equivalent to the actual circuit. This point will be demonstrated in later chapters.

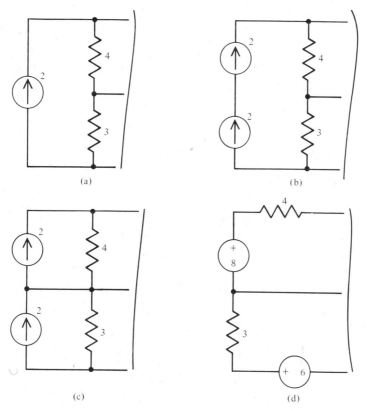

(a)

(b)

(c)

(d)

FIGURE 1–20. Example of compound transformation of a current source to voltage sources.

ACTIVE NETWORK COMPONENTS

The existence of amplifiers, oscillators, radios, TV sets, and all the other electronic equipment used in our world depends on the use of special, active network components. It is not possible to construct this type of equipment from passive components alone. Active components include discrete devices such as transistors, diodes, field effect transistors (FETs), and vacuum tubes, and integrated circuits such as operational amplifiers. The relationship between the voltage across and current through these devices is usually *nonlinear* and they are usually *not* two terminal devices. This means that it is very difficult to solve directly network problems which contain these elements. Solution of networks containing these devices depends on using *linear analytic models* of the networks constructed from standard passive components and special types of sources called *variable dependent or controlled sources*. In this section we study some of the characteristics of active network component models.

1-4.1 Linear Models

The first problem to be considered in using active devices is the inherent nonlinearity exhibited with respect to voltage and current. Consider the diode as an example. The relationship between current through the diode and voltage drop is shown in Fig. 1–21 for a typical unit. The nonlinearity is evident from the fact that the curve is clearly *not a straight line*. In general, a current–voltage (i–v) relationship is *linear* if the equation relating i and v has *both* voltage and

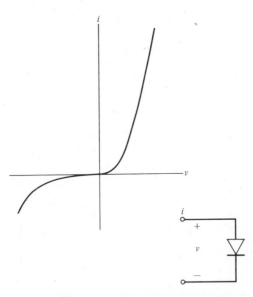

FIGURE 1–21. The *i–v* relationship for a diode is nonlinear.

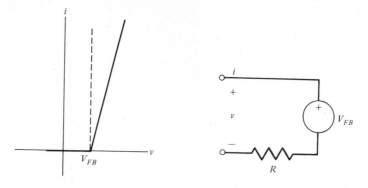

FIGURE 1–22. Linear model of a diode in the forward conducting state.

current *alone and to the first power.* Of course, a resistor is linear, since Ohm's law clearly satisfies this requirement and a graph of current versus voltage for a resistor results in a straight line. A capacitor is also linear, even though a differential relationship exists between voltage and current, as shown in Eq. (1-6), since i and v occur in the first power. In general, it is very difficult to solve network problems if any of the elements are nonlinear. In this text we consider only methods of solving linear networks. In order to solve networks that do contain nonlinear elements, it will be necessary to *approximate* the response of the device with a *linear* model. For example, if a diode is to be used in the forward-biased mode, then the curve, over some range of current and voltage, is approximately linear. It is then possible to use a linear model such as that shown in Fig. 1–22. In this model the resistance represents the small forward resistance of the diode and the voltage source the small forward voltage drop before conduction begins.

1-4.2 Controlled Sources

Many of the active components employed in networks are nonlinear in the i–v characteristic. This is not what makes the devices useful. Their usefulness comes from the fact that they can transform energy in the network from one form to another. For example, an amplifier can transform dc power supply energy into an increase in the energy of the input signal. It is actually a transistor or other active component in the amplifier which performs the transformation. When a linear model of these devices is developed, it is necessary to include something to provide this energy transformation. The *controlled source* is the object in network analysis that models this property. These are also called *variable dependent sources* or simply *dependent sources*.

The network in Fig. 1–23 contains a normal voltage source and a *current-controlled voltage source* (ICVS). The normal voltage source has a value of 10 V with the polarity shown. The ICVS, on the other hand, has a magnitude which is *15 times the current* through the 10-Ω resistor. So its magnitude and polarity are determined by the magnitude and polarity of a current in the net-

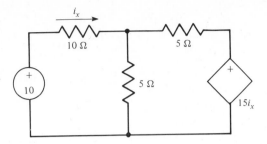

FIGURE 1–23. Network with a controlled voltage source.

work. You should remember that the controlled sources are not *real* in the sense of an actual source but result from *models* of active components. In Chapter 2 you will learn how to solve networks with controlled sources such as that given in Fig. 1–23.

There are four types of controlled sources:

1. Current-Controlled Voltage Source (ICVS). In Fig. 1–24a this is shown to be a voltage source whose magnitude and polarity are dependent on some current magnitude and direction in the network.

2. Current-Controlled Current Source (ICIS). Figure 1–24b shows that this is a current source whose magnitude and direction are controlled by the magnitude and direction of some other current in the network.

3. Voltage-Controlled Current Source (VCIS). This is shown in Fig. 1–24c to be a current source whose magnitude and direction are dependent on the magnitude and polarity of some voltage in the network.

4. Voltage-Controlled Voltage Source (VCVS). In Fig. 1–24d we see this to be a voltage source whose magnitude and polarity are dependent on the magnitude and polarity of some voltage drop in the network.

Note that these controlled sources can be transformed from voltage sources to current sources, and vice versa, just as for real sources, using the transformation process given in Section 1-3.3.

In general, there are many different models of active components. The particular model used depends on the circumstances under which the device is used. Figure 1–25 shows one model of a bipolar transistor which is often used.

The energy that controlled sources deliver to a network actually comes from real sources, but our analytical models do not show this explicitly. If calculations of power requirements are made, the power delivered by controlled sources must be included as coming from real sources connected to the network.

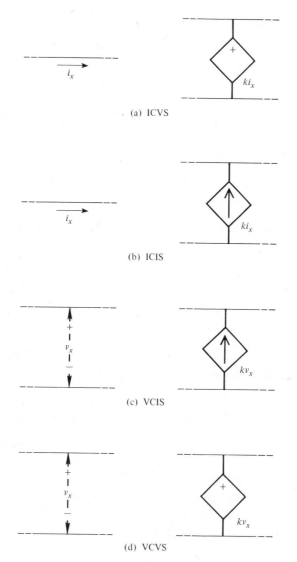

(a) ICVS

(b) ICIS

(c) VCIS

(d) VCVS

FIGURE 1–24. Four types of controlled sources found in networks.

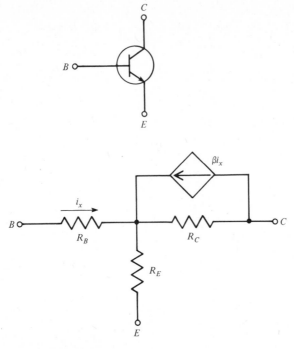

FIGURE 1–25. Controlled sources are used to construct linear models of active devices, such as this transistor.

NETWORK LAWS

In previous sections we have discussed the typical elements found in networks to be solved by network analysis. In this section the two basic laws or principles that are used to solve network problems are reviewed and discussed. A major part of this text is devoted to learning how to apply these laws to various types of networks to obtain a solution. The laws are the result of research and study by many individuals in the history of electrical science, but they have come to be known by the name of the man who formulated them in the most useful way. They are known as *Kirchhoff's voltage law* (KVL) and *Kirchhoff's current law* (KCL).

1-5.1 Kirchhoff's Voltage Law

Kirchhoff's voltage law derives from a consideration of the voltage across the terminals of the elements of a network. The law results from investigations on the conservation of energy in a network and is never violated. KVL is called *instantaneous* because it is valid at every instant of time. First the law will be formally stated and then discussed in terms of a network.

Kirchhoff's Voltage Law. *At every instant of time, the algebraic sum of all voltages across elements in any closed loop of a network will be zero.*

To see the consequences of this law, consider an application to part of some network, such as that shown in Fig. 1–26. Each element has been replaced by a rectangle with the magnitude and polarity of the voltage across that element.

Closed Loop. Notice that several loops have been drawn in the diagram of Fig. 1–26. You can see that a *loop* is any closed connection of elements. We call those loops that do not enclose or encircle any other elements, *elementary loops*. Thus loops 1 and 2 are elementary but loop 3 is not.

Algebraic Sum. When we sum voltages in a loop, all other voltages in the network are ignored, as if the loop were independent of the rest of the

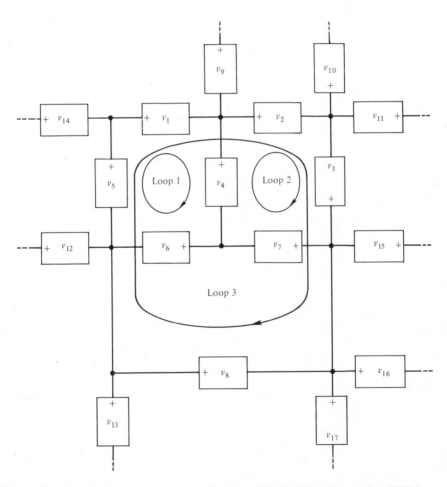

FIGURE 1–26. Part of a network used to define Kirchhoff's voltage law (KVL).

network. The summation must be made by selecting any voltage in the loop and then adding or subtracting the other voltages in the loop according to their polarity with respect to the one selected for the start. For example, if two 6-V sources are in a loop, the summation can be 0 or 12 V, depending on the way they are connected. In applying KVL to a loop the voltages must be combined in accordance with their polarities.

If KVL is applied to the loops of Fig. 1–26, equations for the voltages will result. For loop 1 let us start with v_1; then v_4 will be added since its polarity is in the same sense as v_1 in the loop. Continuing in this way the total sum is

$$v_1 + v_4 + v_6 - v_5 = 0$$

Notice that v_5 has been subtracted since its polarity is in the opposite sense from the other voltages around the loop. Also, it does not really matter which way the algebraic sum runs since the equation above could also have been written

$$v_5 - v_1 - v_4 - v_6 = 0$$

For loops 2 and 3 the corresponding equations are

$$v_2 - v_3 + v_7 - v_4 = 0$$
$$v_1 + v_2 - v_3 - v_8 - v_5 = 0$$

KVL is applied exactly as defined and the result is a set of equations that can be used to find solutions for the currents and voltages in a network.

1-5.2 Application of KVL

Much of the rest of this text is spent learning how KVL can be applied to solve network problems. To illustrate an application of the law, let us use KVL now to derive the familiar rule of how resistors in series combine. Figure 1–27a shows three resistors in series with a voltage source. The four elements form a loop and we have assumed a *loop current* as shown. In Fig. 1–27b the elements have been replaced by rectangles representing the voltages of each device. The polarities are determined by the relation between current direction and resistor voltage drop given in Fig. 1–20. In applying KVL the summation will be started with v_s. This results in the equation

$$v_s - v_1 - v_2 - v_3 = 0 \qquad\qquad (1\text{-}15)$$

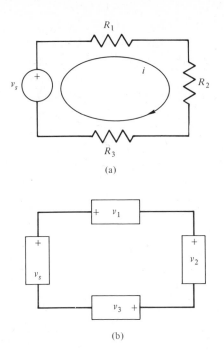

(a)

(b)

FIGURE 1–27. Network for using KVL to find the series combination rule of resistors.

From a mathematical point of view the result is a problem with three unknowns and one equation, an unsolvable set. But from our knowledge of resistors we know that each voltage can be written in terms of the current, using Ohm's law. In this case Eq. (1-15) beomes

$$v_s - iR_1 - iR_2 - iR_3 = 0$$

Now the problem is reduced to one equation in one unknown and is then solvable. Simplifying the expression above gives

$$v_s - (R_1 + R_2 + R_3)i = 0$$

or

$$v_s = (R_1 + R_2 + R_3)i \qquad (1\text{-}16)$$

But Eq. (1-16) has the same form as Ohm's law if the three resistors in series are treated as one resistor of value equal to the sum, $R = R_1 + R_2 + R_3$. The arguments of this problem can obviously be extended to any number of resistors in the series and hence to the general conclusion that resistors in series add.

1-5.3 Kirchhoff's Current Law

Kirchhoff's current law derives from considerations of the conservation of charge in networks. One of the basic postulates of physics is that charge is always conserved in a closed system. This can be extended to networks by noting that if a current carries, say, 150 coulombs per second (C/s) into some interconnection between elements, then 150 C/s must also be carried out. Otherwise, there would be a loss or gain of charges at the interconnection. KCL is an *instantaneous* law because, like KVL, it is valid at every instant of time. The formal statement of this law is:

Kirchhoff's Current Law. *At every instant of time, the algebraic sum of currents entering a node and currents leaving a node is zero.*

A clear picture of this law can be obtained by application to some part of a network, as illustrated in Fig. 1-28. In this case each element has been replaced by a rectangle with the magnitude and direction of the current through the element.

Node.   A node is defined as the point that connects *two* or more elements of a network together. Thus points A, B, and C of Fig. 1–28 are nodes of the network. KCL is defined by currents that enter or leave these nodes.

The algebraic sum is provided by adding those currents that have the same direction relative to the node and subtracting them if they have opposite directions. For convenience the general rule followed is to add currents entering a

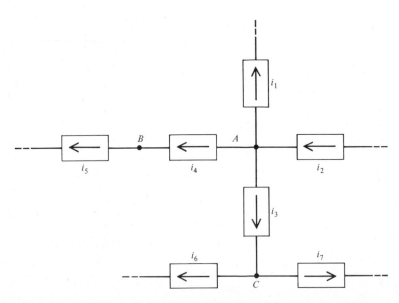

FIGURE 1–28.   Part of a network used to define Kirchhoff's current law (KCL).

node and subtract currents leaving a node. Thus for node A of Fig. 1–28, the equation becomes

$$-i_1 + i_2 - i_3 - i_4 = 0 \tag{1-17}$$

The order of these terms is not important. For nodes B and C we get the equations

$$i_4 - i_5 = 0 \tag{1-18}$$
$$i_3 - i_7 - i_6 = 0 \tag{1-19}$$

In Chapter 2 you will learn procedures on how sets of equations such as these can be used to construct solutions of a network.

1-5.4 Application of KCL

In order to provide an example of the application of KCL, the familiar parallel combination rule for capacitors will be derived. In Fig. 1–29a a circuit contains a current source and three capacitors in parallel. Figure 1–29b shows the same network with each element replaced by a rectangle and an assigned current magnitude and direction. Note that there are two nodes, labeled A and B in the figure. Clearly, the equation for nodes A and B will be *identical* since the same

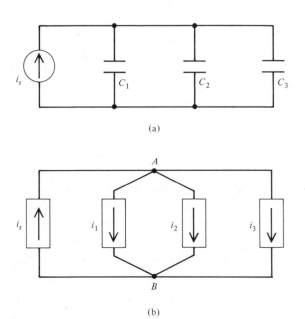

(a)

(b)

FIGURE 1–29. Network for using KCL to find the parallel combination rule for capacitors.

currents are involved at each node. There would be no point in writing equations for each node using KCL since the equations would be the same. For node A the equation from KCL is

$$i_s - i_1 - i_2 - i_3 = 0 \qquad (1\text{-}20)$$

From a mathematical point of view this is one equation in three unknowns (assuming that i_s is known), and therefore the problem cannot be solved. However, there can be only one voltage between nodes A and B, so the same voltage exists across all three capacitors. Furthermore, Eq. (1-6) shows how the voltage across a capacitor is related to the current through the capacitor. If v is the voltage from node A to node B (and the voltage across each capacitor), Eq. (1-20) can be written

$$i_s - C_1 \frac{dv}{dt} - C_2 \frac{dv}{dt} - C_3 \frac{dv}{dt} = 0 \qquad (1\text{-}21)$$

Now Eq. (1-21) can be simplified by factoring dv/dt and written in the form

$$i_s = (C_1 + C_2 + C_3) \frac{dv}{dt} \qquad (1\text{-}22)$$

Equation (1-22) shows that the parallel combination of capacitors in Fig. 1–29a combine in such a way that they act as one capacitor whose value is the sum of the three. This can be easily generalized to any number of capacitors in parallel.

1-6

TIME-DEPENDENT SIGNALS

In many respects network analysis is a study of the variation of currents and voltages in time. Solving a network to determine what functions of time describe the network variables is really only half the problem. The rest is to interpret the solutions to deduce the consequences of the network electrical behavior. This means that it is very important that you have a clear concept of the nature of the kinds of time variations that can occur. In this section the various kinds of time variations that commonly occur in electrical networks are reviewed and discussed.

1-6.1   Time Response

The *time response* of a network can be considered in three distinct categories. These categories describe the general behavior of the network variables in time.

Transient Response.   It is very common in the solution of network variables to find a variation in time which continues for only a finite length of time. For example, the solution may show the presence of an oscillation of voltage, but with decreasing amplitude, so that after a certain length of time the oscillation is no longer present. This type of "short-lived" time response is called *transient behavior*. Nearly all electrical networks exhibit some transient behavior, usually after a "jar" to the system, such as closing some switch or inputting a rapidly changing signal such as a square wave. It is important to be able to recognize transient behavior in a solution.

Steady-State Response.   A solution in network analysis will predict the electrical behavior for all time, or until some change occurs which creates new problem conditions. Part of such a solution will be time variations which persist, without diminishing amplitude, for the duration of the solution. This is the steady-state solution. The steady-state solution is what is left after all transients have decayed to zero effect.

Unstable Response.   In some cases the result of network analysis will be a voltage or current which exhibits an amplitude that *grows without limit*. This is an example of unstable response. Now, of course, the growth in a real circuit will be limited by something since a voltage, say, could not simply grow without limit; something would eventually break down and terminate the growth. Analytically, what terminates the growth is a departure of the network from the conditions under which it was solved (i.e., a new problem, usually nonlinear). Thus a voltage oscillation may increase in amplitude until the saturation of a transistor occurs. Such saturation is a nonlinear effect which would not be covered by the linear model used to set up the problem.

1-6.2   Steady-State Functions

In this section several common types of steady-state response functions are reviewed. You should learn to recognize these functions and be able to visualize the type of time response that each provides.

DC Level.   Often the result of network analysis is the deduction that some voltage or current has a constant value in time. This is called a *dc* (direct-current) signal, which, regardless of the source of the term, can refer to currents or voltages. In equation form, a dc voltage would be written

$$v(t) = V_0 \qquad (1\text{-}23)$$

where V_0 is the value of constant voltage. The implication of a dc signal is that the constant value has existed since the network problem was initiated and will last for as long as the conditions of the problem persist.

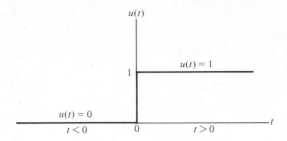

FIGURE 1–30. Plot of the step function u(t).

Step Function.   In many network problems there is a condition that *starts* the problem, such as the closing of a switch to provide power or an input signal. The step function is a signal that can be used mathematically to indicate such an initiation of the problem. In some respects a step function determines when the problem starts (i.e., establishes $t = 0$). The step function, denoted by $u(t)$, has a value of 0 for all $t < 0$ and 1 for all $t > 0$:

$$u(t) = \begin{cases} 0 & \text{for } t < 0 \\ 1 & \text{for } t > 0 \end{cases} \tag{1-24}$$

A plot of $u(t)$ is given in Fig. 1–30. A step function can be used to "close a switch" analytically. For example, application of a 12-V supply voltage to a network at $t = 0$ could be represented by a signal, $12u(t)$. This would be 0 for $t < 0$ and 12 for $t > 0$. (The step function is undefined at $t = 0$.)

A step function can be shifted in time to create a delay or even an advance reaction of some signal. This can be done by specification of the argument of the step function. Thus the function $u(t - a)$ will be 0 for all $t - a < 0$ or $t < a$ and 1 for all $t - a > 0$ or $t > a$. The plot of this function is shown in Fig. 1–31.

Sinusoidal Oscillation.   One of the most common signals in electricity and electronics is derived from the mathematical sine or cosine functions. These

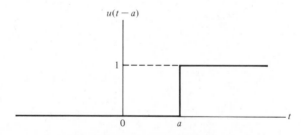

FIGURE 1–31. Displaced step function u(t − a).

functions describe a very characteristic oscillation of current or voltage which results from certain types of networks. Using the sine function, a typical voltage solution would have the form

$$v(t) = V_0 \sin(\omega t) \qquad (1\text{-}25)$$

where

$$V_0 = \text{amplitude}$$

$$\omega = \text{angular frequency, rad/s}$$

The angular frequency, ω, is related to the actual signal frequency in hertz (Hz) by a constant relationship,

$$\omega = 2\pi f \qquad (1\text{-}26)$$

where f is the frequency in hertz. The period of the oscillation is, of course, given by $T = 1/f$.

Figure 1–32 shows a plot of a sine oscillation such as that given by Eq. (1-25) and identifies the amplitude and period. Note that any sinusoidal oscillation which starts at zero when $t = 0$ must be described by the sine function, with no phase shift.

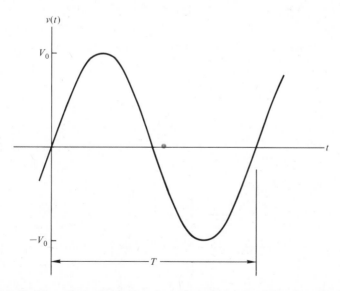

FIGURE 1–32.   Plot of the sine function sin (ωt), where $T = 2\pi/\omega$.

EXAMPLE 1-10

A current is given by $i(t) = 2 \sin (125t)$. Find the peak value of the current, the period, and the frequency.

SOLUTION

The peak value is just the amplitude, as given by Eq. (1-25). Thus the peak current is simply 2 A. By comparison with Eq. (1-25) it is clear that $\omega = 125$ rad/s. Then from Eq. (1-26) the frequency is $f = \omega/2\pi = 125/(2)(3.14159) = 19.89$ Hz. Then the period is $1/f$ or $T = 0.05$ s. ■

EXAMPLE 1-11

A periodic function starts from zero at $t = 0$ and varies from $+13$ to -13 with a frequency of 67 Hz. Find the function to describe this signal.

SOLUTION

Since the function starts from zero at $t = 0$ it must be a sine function. The amplitude is just the peak value and thus $V_0 = 13$. The angular frequency is found from Eq. (1-26):

$$\omega = 2\pi f = 2(3.14159)(67) \approx 421 \text{ rad/s}$$

Thus the equation is

$$v(t) = 13 \sin (421t)$$
■

The cosine function has exactly the same form as the sine function except that it is shifted in phase by $\pi/2$ radians (90°). This means that the cosine function starts at the maximum of the signal (i.e., the peak value). The analytic expression of a cosine function would be

$$v(t) = V_0 \cos (\omega t) = V_0 \sin \left(\omega t + \frac{\pi}{2} \right) \qquad (1\text{-}27)$$

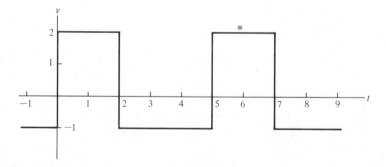

FIGURE 1–33. Plot of a nonsymmetric square wave.

A general sinusoidal type of oscillation that is not starting from the peak or from zero at $t = 0$ can be described by one of three equivalent expressions:

$$v(t) = V_1 \sin(\omega t) + V_2 \cos(\omega t)$$

or

$$v(t) = V_a \sin(\omega t + \phi)$$

or

$$v(t) = V_b \cos(\omega t + \psi)$$

where the relationships between amplitudes and phases are determined from trigonometry.

Rectangular Waves. The rectangular wave is a very common signal in electrical networks. This signal is characterized by having only two levels and of making periodic, instantaneous transitions between these levels. Figure 1–33 shows a rectangular wave which makes transitions between the levels of $+2$ and -1 with a period of 5 s. Note that the level is at -1 for a longer period of time than it is at $+2$. In some cases rectangular waves have symmetry, such as equal times at each level (time symmetry) and equal level magnitudes (level symmetry).

Triangular Waves. There are many types of signals which involve linear ramps. Figure 1–34 shows two types of triangular waves that are encountered in network analysis. The signal in Fig. 1–34a is called a *sawtooth* wave. It rises linearly (ramps) from zero to a peak value of 20 in a period of 2 s and then drops instantaneously back to zero level to begin a rise again. The signal in Fig. 1–34b ramps symmetrically about the level axis from 10 to -10 in a period of 4 s.

1-6.3 Transient Functions

There are several types of transient functions which commonly appear in the solution of network problems. It is important to be able to recognize these functions and understand the significance of their temporal behavior.

Exponential Decay. One of the most common transient signals that occurs in electrical networks is a simple, smooth decay to zero from some starting value. This is described by a characteristic equation and is called exponential

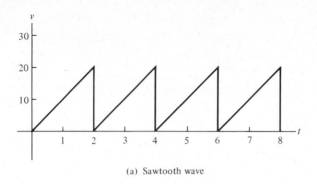

(a) Sawtooth wave

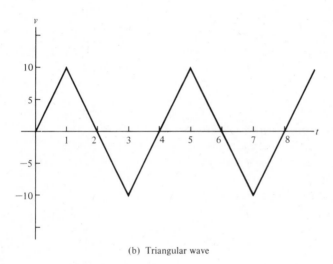

(b) Triangular wave

FIGURE 1–34.   Plots of the sawtooth and triangular waves.

decay. One common example is the discharge of a capacitor through a resistor. The general functional form is

$$v(t) = V_0 e^{-t/\tau} \tag{1-28}$$

where

$$V_0 = \text{initial amplitude}$$

$$\tau = \text{time constant,}$$

The *time constant* is a measure of the time that it takes for the signal to decay away to zero. The time constant is also known as the decay time. In some cases it is written as the inverse, which is called the *decay constant*, $\alpha = 1/\tau$. Figure 1–35 shows a plot of a typical exponential decay curve.

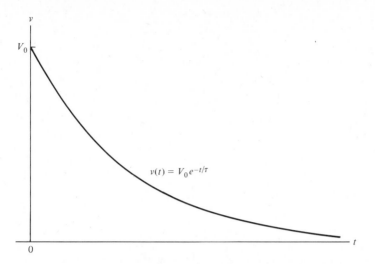

FIGURE 1–35. Plot of the exponential decay transient.

The larger the time constant, the longer the signal will take to die away. A convenient measure of the time the signal will persist is given by knowing that the signal will fall to about 37% of V_0 in a time equal to one time constant,

$$v(\tau) = V_0 e^{-\tau/\tau} = V_0 e^{-1} = 0.368V_0$$

A current or voltage described by a function such as Eq. (1-28) will never actually reach zero since the function is asymptotic to zero. Practically, it is common to state that the transient has decayed to zero after a time equal to *five time constants*.

EXAMPLE 1-12
An exponential decay voltage has a value of 15 V at $t = 0$ and 6.0 V at $t = 2$ s. Find the time constant.

SOLUTION
From Eq. (1-28) it is clear that $V_0 = 15$. Then at $t = 2$ this equation becomes

$$15e^{-2/\tau} = 6.0$$

$$e^{-2/\tau} = 0.4$$

Taking logarithms of both sides gives

$$\ln(e^{-2/\tau}) = \ln(0.4)$$

$$-\frac{2}{\tau} = -0.916$$

$$\tau = 2.18 \text{ s} \qquad \blacksquare$$

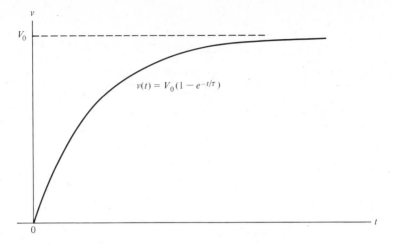

FIGURE 1–36. Plot of an exponential transient startup signal.

Exponential Startup. Another common transient is that which occurs when a signal changes in a characteristic, smooth fashion from one value to another. This is really just a modification or application of the exponential decay curve discussed above, but its common appearance suggests that it be presented separately. In the common case when the change is from zero to some value, the function has the form

$$v(t) = V_0(1 - e^{-t/\tau}) \tag{1-29}$$

where the terms have been defined above. The characteristic shape of this curve is shown in the plot of Fig. 1–36. A common interpretation of this type of signal is to say that in one time constant, $t = \tau$, the function has accomplished about 63% of the change:

$$V(\tau) = V_0(1 - e^{-1}) = 0.632V_0$$

Oscillating Transient. In many cases a transient occurs which exhibits an oscillation as it decays to zero. In this case the signal does not decay smoothly to zero as an exponential decay, but oscillates in value as the amplitude of the oscillation decays. This is sometimes called *ringing*. This type of transient, shown in Fig. 1–37, can be described by a sinusoidal function *and* an exponential decay. An example would be

$$v(t) = V_0 e^{-t/\tau} \sin(\omega t + \phi) \tag{1-30}$$

where all the terms have been defined earlier.

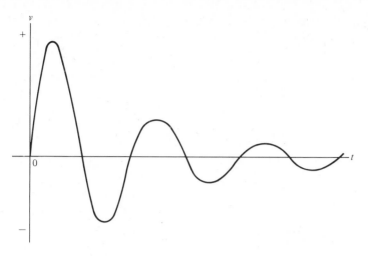

FIGURE 1–37. Plot of an oscillating transient which is also called a ringing signal.

EXAMPLE 1-13

A ringing transient has the form

$$v(t) = 12e^{-t/4} \sin (8t - 0.5)$$

Find the frequency of oscillation in hertz and the voltage at 0.5 s.

SOLUTION

The frequency is found from the fact that $\omega = 8$ rad/s. Then $f = \omega/2\pi = 8/2\pi = 1.27$ Hz. The voltage at 0.5 s is

$$v(0.5) = 12e^{-0.5/4} \sin (4 - 0.5)$$

$$= 12(0.88)(-0.35)$$

$$= -3.7 \text{ V} \qquad \blacksquare$$

Impulse. The impulse, although not really a function at all, is important to the study of network analysis. The impulse or delta function is really the result of a limiting process on the derivative of a step function. It is an attempt to represent the effect of a very short lived electrical "blow" on a circuit. Such effects can be substantial. From a physical point of view the impulse can be understood by analogy to a bat striking a ball. The conditions before contact are described physically by things like the velocity of the bat and the ball and perhaps the energy content of each. No attempt is made to describe, exactly, the blow of the bat against the ball. Rather, conditions after the impact are described in terms of the velocity and energy of the bat and ball. This is because highly complicated and often nonlinear things go on during the very short time during which actual impact occurs. So it is in some cases in network analysis. Particularly when capacitors and inductors are involved in networks, it is found that the energy stored in these elements can cause impulse-like reactions in a network when they are connected to the network.

It turns out not to be necessary to use the value of the impulse, so whether it is really infinite or not is not important. Analytically, the impulse can be described as a pulse at $t = 0$ whose width is decreased by a limiting process while the area under the curve remains unity. This is shown in Fig. 1–38. Since the width is W and the area is unity, the height must be $1/W$. So when the width goes to zero, the height goes to infinity. The symbol of the impulse function is $\delta(t)$. In principle it can be defined by

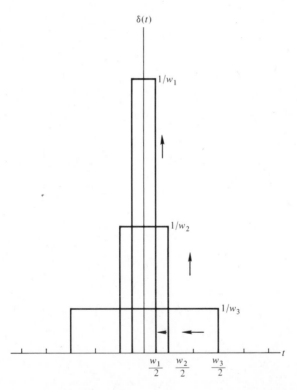

FIGURE 1–38. The impulse or delta function is the limit of a signal whose width decreases as the height increases, in inverse proportion.

INTRODUCTION TO NETWORK ANALYSIS

$$\delta(t) = \frac{du(t)}{dt} \qquad (1\text{-}31)$$

where the constraint is

$$\int_{-\infty}^{\infty} \delta(t) \, dt = 1 \qquad (1\text{-}32)$$

Like the step function, the impulse can be delayed in time. A delta function written as $\delta(t - a)$ is an impulse that occurs at $t = a$ instead of $t = 0$.

1-6.4 Computer Usage

Often a digital computer is of value in the interpretation and plotting of functions that result from network analysis. In general, the program library of a computer installation will have a plotting routine. In this case it is of value to input particularly complex functions into such a routine and deduce the behavior from visual observation of the resulting plots.

EXAMPLE 1-14

The following function describes a network voltage. Plot this function and determine the maximum voltage and when it occurs, and the length of time between when the voltage first exceeds ± 5 V and last drops below ± 5 V.

$$v(t) = 10e^{-(t-4)^2} \sin(12t)$$

SOLUTION

The voltage signal was plotted using a computer routine with a large time scale to see the overall behavior of the function. Figure 1–39a shows that the voltage is an oscillating pulse centered on 4 s. Next a more detailed plot was made, as shown in Fig. 1–39b, so that the questions could be answered with some degree of accuracy. From this plot the positive maximum is seen to occur at 3.8 s with a value of 25 increments and the negative at 4.05 s with a -25 increment peak. The printout indicates that each increment is 0.397 V. Thus the peak voltages are $(\pm 25)(0.397) = \pm 9.93$ V. To find the ± 5-V times, lines are drawn at $+5$ (12.5 increments) and -5 V (-12.5 increments). This is shown on the graph, which also shows that the curve exceeds this value at 3.25 s and drops below at 4.65 s, so the total time is 1.4 s. All of these values are limited by the resolution of the plot. For greater accuracy it would be necessary to construct plots with better resolution. ■

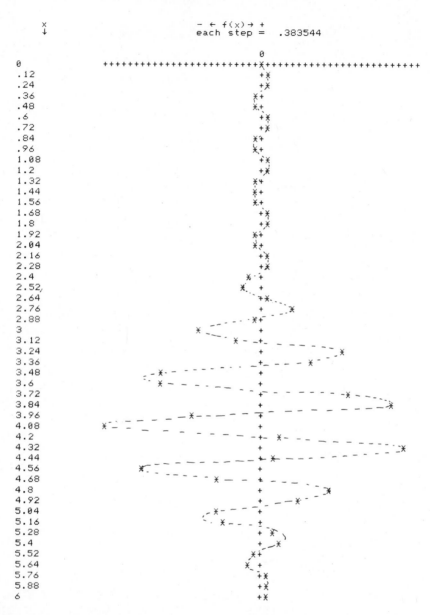

FIGURE 1–39a. Computer plot of an oscillating pulse transient: large-scale plot to show overall behavior.

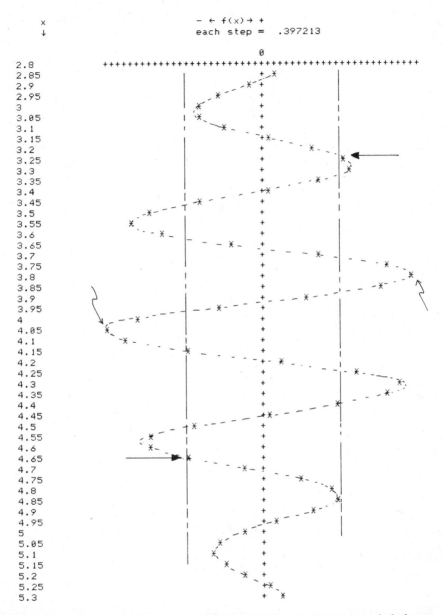

FIGURE 1—39b. Computer plot of an oscillating pulse transient: expanded plot to show details of the pulse.

SUMMARY

The primary objective of this chapter was to review basic network concepts, definitions, and terms. The following important topics were discussed:

1. The network was defined as an assemblage of elements and network variables were defined as the voltage across and the currents through these elements.
2. For the passive network components—resistors, capacitors, inductors, and transformers—the current versus voltage relations were given as well as the relation for power dissipation or energy storage.
3. The properties of ideal network voltage sources and current sources were defined. The transformation rules for converting voltage sources to current sources, and vice versa, were given.
4. The concept of controlled sources was given, showing that these elements result from modeling active network components.
5. Formal definitions were given for the two basic circuit laws on which network analysis is based: Kirchhoff's voltage law and Kirchhoff's current law.
6. A summary was given of the typical steady-state time functions encountered in network analysis. This included the dc level, step function, sinusoidal oscillation, and nonsinusoidal oscillations.
7. The common network transient functions were defined. The functions presented were the exponential decay, exponential startup, oscillating transient, and impulse or delta function.

PROBLEMS

1-1 A network element has a current of $i(t) = 5 \cos (2t)$ and a voltage of $v(t) = 10 \sin (2t)$. Plot the power versus time. Describe the plot in terms of energy delivered to or from the element.

1-2 Plot the energy associated with the element of Problem 1–1.

1-3 A network element has a current of $10e^{-4t}$ and a voltage drop of $5(1 - e^{-4t})$. Plot the power and energy versus time.

1-4 Use the series and parallel rules of combining resistors to find the voltage drop, current, and power associated with each resistor in Fig. 1–40.

1-5 A 4-F capacitor has a voltage drop given by the equation $v(t) = 3t^2 + 5t - 8$ V. What is the current through the capacitor?

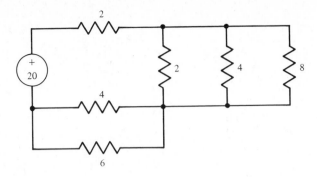

FIGURE 1–40. Network for Problem 1–4.

1–6 A capacitor has a voltage drop of $10e^{-5t}$ V. If the capacity is 2 F, plot the energy stored in the capacitor versus time.

1–7 A capacitor is charged by a steady flow of charge equal to 0.65 C/s for 3 min. What energy is stored in the capacitor after this time? What is the voltage across the capacitor? The capacity is 2.3 F.

1–8 An inductor of 3 H has a current of $4e^{-0.2t} \cos(23t)$ A. What is the voltage across the inductor?

1–9 What energy is stored in the inductor of Problem 1–8 at $t = 0$? What energy at $t = 10$ s?

1–10 An ideal transformer with positive connected mutual inductance has windings with $n_1 = 50$ turns and $n_2 = 7500$ turns. If the voltage on winding 1 is $v_1(t) = 2 \sin(5t)$, find $v_2(t)$.

1–11 In Problem 1–10, the current is found to be $i_1(t) = 5 \cos(5t)$. Find $i_2(t)$ and compare the power at the two sets of transformer terminals.

1–12 A transformer has $L_1 = 5$ H, $L_2 = 18$ H, and $M = 13$ H with a negative connection. Find the T model.

1–13 Plot the load power versus load resistance connected to a battery of 10 V and 5.7-Ω internal resistance. At what resistance is the power a maximum?

1–14 Given a voltage source of $25e^{-4t}$ V with a 5-Ω internal resistance, construct the equivalent current source circuit.

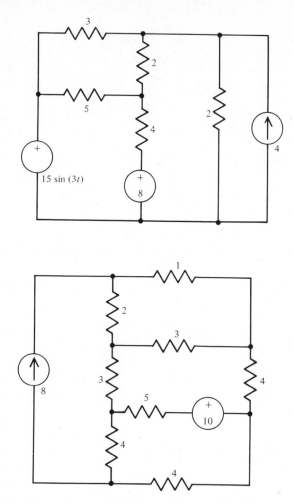

FIGURE 1–41. Network for Problems 1–15 and 1–16.

1–15   Transform all voltage sources to current sources in the networks given in Fig. 1–41 (top). Redraw the network with the current sources.

1–16   Transform all current sources to voltage sources in the networks given in Fig. 1–41 (bottom). Redraw the network with the voltage sources.

1–17   For the controlled source network of Fig. 1–42 find the current, voltage, and power associated with the 100-Ω resistor.

1–18   Repeat Problem 1–17, but transform the controlled current source and 10-Ω resistor into a controlled voltage source and resistor before solving.

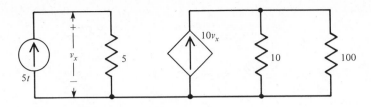

FIGURE 1–42. Network for Problem 1–17.

1–19 Use KVL to find the values of V_1, V_2, and V_3 in Fig. 1–43.

1–20 Use KCL to derive the parallel combination law for resistors.

1–21 Write time equations for each of the voltage signals given in Fig. 1–44.

1–22 Given a current $i(t) = 25(1 - e^{-3t/4})$ A, find the time at which the current reaches 92% of its final value.

1–23 A resistor is connected to a voltage source given by the equation $v(t) = 25e^{-t/2} \sin (15t + 1)$. If the resistance is 20 Ω, find the current and power at 1.5 s and 4.5 s.

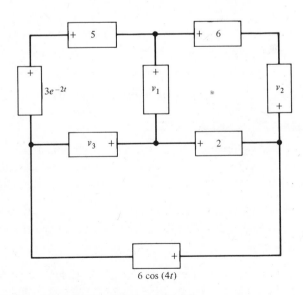

FIGURE 1–43. Network for Problem 1–19.

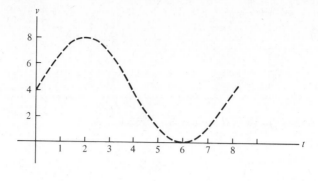

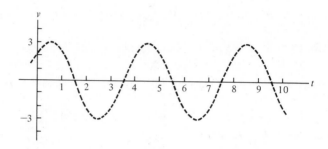

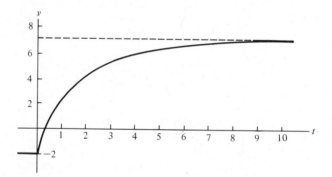

FIGURE 1–44. Illustration for Problem 1–21.

COMPUTER PROBLEMS

1–24   Use a plotting program to plot the power in Problem 1–23. At what time is the power at the maximum, and what is the value? At what times does the power reach a minimum? At what time does the power fall to 25% of the maximum?

1–25   A network element has a current of $i(t) = -5e^{-4t} \cos (14t)$ and a voltage of $7(1-e^{-4t}) \cos (14t)$. Plot the power. What is the maximum power, and when does it occur?

1–26   The current to a 4-H inductor is $5e^{-4t} \sin (20t + 0.8)$ A. Find the voltage across the inductor. Find the power. Plot the current, voltage, power, and energy associated with the inductor. Discuss phase shifts between these quantities.

CHAPTER 2

Solving Resistive Networks

Objectives

The single global objective of this chapter is to give the reader the skill to solve any resistive network. This can be specified in more detail through several subobjectives. In particular, then, after a thorough reading of this chapter and completion of the problems at the end of the chapter, you will be able to:

1. Define the characteristics of resistive networks.
2. Use KVL (loop analysis) or KCL (node analysis) to construct sets of simultaneous equations to solve a network.
3. Define mesh analysis and use to set up the matrix equation for the simultaneous equations to solve a resistive network.
4. Define nodal analysis and use to set up the matrix equation for the simultaneous equations to solve a resistive network.
5. Use Cramer's rule and Laplace expansion of higher-order determinants to solve for the unknowns in a matrix equation.
6. Use mesh and/or nodal analysis techniques to aid in finding the Thévenin and Norton equivalent circuits of a resistive network.

RESISTIVE NETWORKS

There are many networks that occur in electricity and electronics which have only resistors as the passive component elements; that is, they do not have capacitors, inductors, or transformers. These are called *resistive networks*. There is no restriction on the voltage or current sources. Furthermore, the network may contain active devices which have been modeled as controlled sources. Solutions for the voltages and currents in such a network requires the solution of a set of simultaneous, linear equations, but no calculus is involved. One method of finding the equations uses a direct application of KVL and/or KCL to the network. Two other methods, although based on KVL and KCL, use formal procedures to construct matrix representations of the simultaneous equations. These methods have the names *mesh analysis* and *nodal analysis*.

DIRECT SOLUTION BY KVL AND KCL

A resistive network can be solved by using KVL and KCL, in combination with Ohm's law, to set up sufficient equations to solve for all the unknowns. The unknowns will be voltages across components and/or currents through components. Obviously, if we know either all the currents or all the voltages, the problem is completely solved since Ohm's law can be used to find one variable from the other. It is much easier to solve a network if we decide ahead of time that either only voltages or only currents will be found directly. This is determined by which law is used to set up the initial set of equations to be solved, as shown in the following sections.

2-2.1 Solving for Currents Using KVL: Loop Analysis

The use of KVL to solve a network will result in equations for the currents in the network. This method is also called *loop analysis* since the use of KVL means that voltages around a loop of the network wi¹l be summed to zero to obtain an equation. It will then be necessary to use KCL and Ohm's law to simplify these equations. Since it is voltages that are summed, it is helpful if all sources are expressed as voltage sources when applying this law. Therefore, if any current sources are present, they should be transformed to voltage sources using the methods outlined in Chapter 1. This transformation is not *required;* it just makes subsequent analysis a little easier.

Consider the network of Fig. 2–1. The application of KVL to this circuit is accomplished by first writing the network in terms of the voltage across each element. This is shown in Fig. 2–2a, where the polarities of voltage drops across

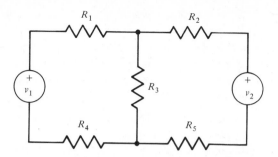

FIGURE 2–1. Network for the application of KVL loop analysis.

the resistors have simply been assigned arbitrarily. Now, by KVL, these voltages are summed for each loop, and since there are three loops there will be three equations.

$$v_1 - v_{R1} - v_{R3} - v_{R4} = 0$$

$$v_{R2} + v_2 - v_{R5} - v_{R3} = 0$$

$$v_1 - v_{R1} - v_{R2} - v_2 + v_{R5} - v_{R4} = 0$$

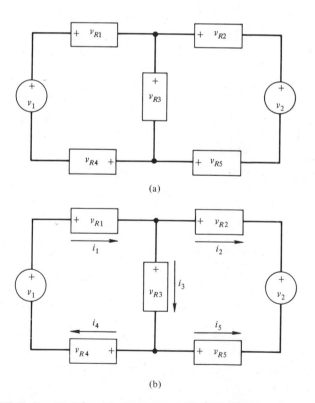

FIGURE 2–2. In (a) each element of Fig. 2–1 has been replaced by the voltage across its terminals. In (b) the appropriate current has been identified.

Notice that the voltages must be combined in accordance with their polarities. Well, this is three equations in *five* unknowns! Clearly, there are an insufficient number of equations. What is needed, then, is more equations or a reduction in the number of unknowns, by using other facts about the variables. This can be done by assigning a current to each element and expressing the voltage as a function of the current. Figure 2–2b shows the same network of voltage drops but with assigned currents. The direction of the currents were determined from the fact that the current must enter the resistor on the side with the positive voltage drop polarity. The equations above can be written in terms of these currents using Ohm's law:

$$v_1 - R_1 i_1 - R_3 i_3 - R_4 i_4 = 0 \qquad (2\text{-}1)$$

$$R_2 i_2 + v_2 - R_5 i_5 - R_3 i_3 = 0 \qquad (2\text{-}2)$$

$$v_1 - R_1 i_1 - R_2 i_2 - v_2 + R_5 i_5 - R_4 i_4 = 0 \qquad (2\text{-}3)$$

The problem seems to have gotten no better since the three equations still have five unknowns, now currents. However, these can be simplified quite easily. It is clear, for example, that $i_1 = i_4$ and that $i_2 = -i_5$ since KCL shows that only one current can flow through a series of elements. Also, from KCL applied to the upper junction of the three resistors, it is found that $i_1 - i_2 - i_3 = 0$, or $i_3 = i_1 - i_2$. These three relations for the currents show that the equations above can now be expressed *entirely* in terms of *two* currents, i_1 and i_2. So now the problem is overspecified because there are three equations and only two unknowns. This will always be the case when applying loop analysis by the method shown here. The way out of this dilemma is simple; we just use two of the equations. The usual choice is the equations that come from the *elementary loops* of the network. The elementary loops, as defined earlier, are those that do not enclose any network elements. It turns out that if the elementary loops are used, the correct number of equations will always result. Notice that the two elementary loop currents actually form loops, i_1 in the left part of the circuit and i_2 in the right part of the circuit. Both of these loop currents pass through the shared resistor, R_3. Since the two loop currents are sufficient to solve the network, only Eqs. (2-1) and (2-2) need to be used to solve the problem network.

When Eqs. (2-1) and (2-2) are combined with the relations for the currents given in the previous paragraphs, the resulting set of simultaneous equations are

$$(R_1 + R_3 + R_4)i_1 - R_3 i_2 = v_1 \qquad (2\text{-}4)$$

$$-R_3 i_1 + (R_2 + R_3 + R_5)i_2 = -v_2 \qquad (2\text{-}5)$$

The problem is now reduced to finding the solution of a set of simultaneous equations. It is a math problem and the network analysis is essentially finished.

If the number of equations is small, the method of substitution can be used. For large sets of equations, Cramer's rule, which will be reviewed later in this chapter, can be used, or we can use computer programs that solve simultaneous equations or determinants. The following example illustrates the method of substitution.

EXAMPLE 2-1

For the circuit of Fig. 2–1, as developed above, find i_1 and i_2 if the component values are: $v_1 = 5$ V, $v_2 = 3$ V, $R_1 = 2$ Ω, $R_2 = 5$ Ω, $R_3 = 4$ Ω, $R_4 = 3$ Ω, and $R_5 = 6$ Ω. Find the voltage across the 3-Ω resistor.

SOLUTION

In this case the equations have already been found for i_1 and i_2, as Eqs. (2-4) and (2-5). Using the given values of resistors and sources gives the set

$$9i_1 - 4i_2 = 5$$

$$-4i_1 + 15i_2 = -3$$

Solving the first equation for i_1 gives

$$i_1 = \frac{5}{9} + \frac{4i_2}{9}$$

This result is substituted into the second equation,

$$-4\left(\frac{5}{9} + \frac{4i_2}{9}\right) + 15i_2 = -3$$

This equation can now easily be solved for i_2,

$$i_2 = -0.059 \text{ A}$$

and i_1 is then found from the first equation,

$$i_1 = 0.529 \text{ A}$$

The fact that i_2 came out negative means that the actual current direction is opposite from our assumed direction through the second loop. To find the voltage across R_3, it will be necessary to multiply the *net* current through the resistor by the resistance. The net current is given by $i_3 = i_1 - i_2 = 0.529 - (-0.059) = 0.588$ A. Then the voltage is $v = 4(0.588) = 2.35$ V. The positive is on the top of the resistor since that is the direction from which the net current comes. ■

The method of substitution demonstrated by this example will work for any number of simultaneous equations, although it is clear that the method would become cumbersome for very many equations. In general, for a network with four or more elementary loops, and therefore four or more equations, a technique such as Cramer's rule will be much easier to use. Mesh analysis will show how the set of simultaneous equations can be set up, in matrix formulation, by a much easier process than this direct application of KVL.

2-2.2 Solving for Voltages Using KCL: Node Analysis

The use of KCL for setting up the equations for a network will result in voltages as the unknowns. Then KVL and Ohm's law are used to simplify the resulting equations. Since KCL is applied by adding currents at nodes, the sources should be expressed as current sources. If there are any voltage sources in the network, it will be helpful to transform them to current sources. This transformation is not required; it just makes the solution a little easier to set up.

Consider the network of Fig. 2–3, where the nodes have already been identified and labeled 1, 2, 3, and 4. The first step in applying KCL to a network is to assign a current to each element. In Fig. 2–4 each element has been replaced by such an assumed current and node voltages have been assigned using node 4 as the reference. The direction of the currents are chosen arbitrarily. An application of KCL now specifies that the sum of the currents at each node is zero. From this the following equations result:

$$i_1 - i_{R1} = 0$$

$$i_{R1} - i_{R3} - i_{R2} = 0$$

$$i_{R2} - i_2 + i_{R4} = 0$$

This set of three simultaneous equations involves four unknowns and is therefore not solvable. To obtain a solvable set we will reduce the number of unknowns.

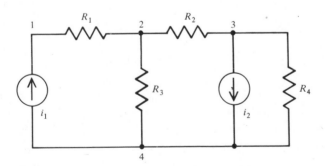

FIGURE 2–3. This network will be used to demonstrate the application of KCL to solve a network.

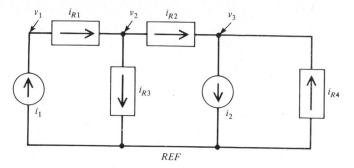

FIGURE 2–4. Each element of Fig. 2–3 has been replaced by a current and node voltages have been assigned.

To do this, the currents in the equations are expressed in terms of the node voltages. Thus there will be three unknowns, v_1, v_2, and v_3. Ohm's law can be used to express each unknown current in the equations above in terms of these three voltages. In this manner the result is three equations in three unknowns, a solvable set. In expressing the currents in terms of the node voltages, the direction of the current determines which voltage is assumed to be at a higher potential, since current flows from high to low potential. Thus for i_{R1} the direction is from node 1 to node 2, so the assumption is that v_1 is higher than v_2 and the current must be written

$$i_{R1} = \frac{v_1 - v_2}{R_1}$$

By the same kind of reasoning the remaining currents are

$$i_{R2} = \frac{v_2 - v_3}{R_2}$$

$$i_{R3} = \frac{v_2}{R_3}$$

$$i_{R4} = -\frac{v_3}{R_4}$$

Notice that the last current is negative because the assumed current direction was from the reference to node 3. When these relations for the currents are substituted into the KCL equations, the result is

$$\frac{1}{R_1}v_1 \qquad -\frac{1}{R_1}v_2 \qquad\qquad = i_1 \qquad\qquad (2\text{-}6)$$

$$ -\frac{1}{R_1} v_1 + \left(\frac{1}{R_1} + \frac{1}{R_2} + \frac{1}{R_3} \right) v_2 - \frac{1}{R_2} v_3 = 0 \qquad (2\text{-}7) $$

$$ -\frac{1}{R_2} v_2 + \left(\frac{1}{R_2} + \frac{1}{R_4} \right) v_3 = -i_2 \qquad (2\text{-}8) $$

Equations (2-6), (2-7), and (2-8) are the simultaneous set of equations that will solve the network of Fig. 2–3.

EXAMPLE 2-2

Find the node voltages and the current through R_2 for the network in Fig. 2–3. The elements have the following values: $i_1 = 2$ A, $i_2 = 4$ A, $R_1 = 10\ \Omega$, $R_2 = 5\ \Omega$, $R_3 = 2\ \Omega$, and $R_4 = 4\ \Omega$.

SOLUTION

This problem is solved by using the simultaneous equations given above with the values assigned by the problem statement. This provides the set

$$ 0.1v_1 - 0.1v_2 \qquad\qquad = \quad 2 $$

$$ -0.1v_1 + 0.8v_2 - 0.2\ v_3 = \quad 0 $$

$$ - 0.2v_2 + 0.45v_3 = -4 $$

The first equation can be easily solved for v_1:

$$ v_1 = 20 + v_2 $$

If this is substituted into the next two equations, the result is two equations for v_2 and v_3:

$$ 0.7v_2 - 0.20v_3 = \quad 2 $$

$$ -0.2v_2 + 0.45v_3 = -4 $$

The first equation gives $v_3 = 3.5v_2 - 10$. If this is substituted into the last equation, the result is

$$ -0.2v_2 + 0.45(3.5v_2 - 10) = -4 $$

So from this equation, $v_2 = 0.36$ V. When this is substituted back into the previous relations for the other voltages, the results are

$$ v_1 = 20.36\ \text{V} $$

$$ v_3 = -8.73\ \text{V} $$

The current through R_2 is found from Ohm's law since the voltage drop across the resistor is known.

$$i_{R2} = \frac{v_2 - v_3}{R_2}$$

$$= \frac{0.36 + 8.73}{5} = 1.82 \text{ A} \qquad \blacksquare$$

2-2.3 Summary of the Direct Approach

The methods of solving networks outlined in previous sections can be used to solve any linear resistive network. The examples given were for particularly simple networks, with few elements and few unknowns. When more complicated networks are considered, a number of difficulties of a practical nature occurs. First, the number of unknowns and equations become so large that it is difficult to keep track of the relationships between unknowns. In this case it sometimes becomes difficult to determine enough relationships to produce the required number of equations for the number of unknowns. Second, with many equations the method of substitution becomes unmanageable for solving for the unknowns.

Fortunately, more practical methods have been developed for setting up the simultaneous equations from the network. Using these methods, called mesh and nodal analysis, the simultaneous equations can often be written down directly from the network, by inspection! This reduces the difficulties associated with the first point above. The second difficulty is reduced by using a matrix approach to express the simultaneous equations and then solving these equations using Cramer's rule and the evaluation of higher-order determinants. Before presenting the mesh and nodal methods of setting up simultaneous equations of networks it is necessary that you have a general understanding of matrix algebra. Furthermore, from the practical side of finding the values of unknowns, it is necessary that you know how to evaluate determinants of order greater than 3. A study of these math topics serves now as a prerequisite to learning the advanced network solving methods.

2-3

MATRICES, SIMULTANEOUS EQUATIONS, AND DETERMINANTS

It is very common in science and technology to encounter problems which result in sets of simultaneous equations. In response to this type of problem, special mathematical techniques have been developed to simplify the representation, analysis, and solution of the simultaneous equations. This section presents the essential and practical features of these techniques, which can be applied to the solution of network problems.

2-3.1 Matrix Representation

One of the techniques developed to ease the handling of sets of simultaneous equations started out as simply a more convenient way of writing these equations. This led to the development of a new branch of mathematics called *matrix algebra*. This section has been titled *matrix representation* rather than matrix algebra since the principal use of matrices in this text will be as a convenient way of constructing and representing the simultaneous equations from the network.

Matrix. A matrix is simply an array of numbers for which some *set of rules* specify how the array is constructed. In a sense, then, it is like a system of bookkeeping which keeps track of a set of numbers. The matrix itself has no specific, single value. When a variable name is assigned to a matrix such as, "let **A** be a matrix," it does *not* mean that **A** has some value but rather that **A** is a *symbol* representing an array of numbers. A matrix has a property called *dimension*. The dimension is actually a measure of the complexity of the rules necessary to construct the array of numbers.

A one-dimensional matrix is called a *linear array* because there is only *one rule* to specify the construction of the array. Suppose that it is required to form the inventory of resistors with values of 100 Ω, 2.7 kΩ, 4.7 kΩ, 33 kΩ, and 100 kΩ. The inventory data can be represented as an inventory matrix, **A**, by a single rule which lists the quantity of each value of resistor in stock. If expressed as a *column matrix*, **A** might have the form

$$\mathbf{A} = \begin{pmatrix} 40 \\ 22 \\ 0 \\ 15 \\ 8 \end{pmatrix} \quad \begin{array}{l} \text{(number of 100-}\Omega\text{ resistors)} \\ \text{(number of 2.7-k}\Omega\text{ resistors)} \\ \text{(number of 4.7-k}\Omega\text{ resistors)} \\ \text{(number of 33-k}\Omega\text{ resistors)} \\ \text{(number of 100-k}\Omega\text{ resistors)} \end{array}$$

Obviously, the rule by which the array is constructed is that the inventory is arranged as the stock, from the top down in increasing values of resistance. This matrix has five *rows*, each with one number. Then row 1 has the number 40, which is the number of 100-Ω resistors, row 2 is the number of 2.7-kΩ resistors, and so on. This is called a 5 × 1 (five by one) *column matrix* since it has five rows and one column.

The same inventory data could be represented by a *row* matrix by slightly changing the rule for construction. Then the matrix would be a 1 × 5 row matrix,

$$\mathbf{B} = (40 \quad 22 \quad 0 \quad 15 \quad 8)$$

A *two-dimensional matrix* is one which has more than one row and more than one column. In this case there is a second rule on how the array is constructed.

Suppose in the inventory above that the stock of resistors is to be further distinguished by the power rating of each resistor. For example, the resistors may be distributed in $\frac{1}{4}$-W, $\frac{1}{2}$-W, and 1-W ratings. If the rows represent values of resistance, as in matrix $\mathbf{A}$ above, the columns will now be used to indicate the power rating. Suppose that the leftmost column (column 1) is $\frac{1}{4}$ W, column 2 is $\frac{1}{2}$ W, and column 3 is 1 W. A typical inventory matrix might now be

$$\mathbf{C} = \begin{pmatrix} 10 & 25 & 5 \\ 0 & 12 & 10 \\ 0 & 0 & 2 \\ 5 & 2 & 8 \\ 4 & 4 & 0 \end{pmatrix}$$

According to the rules for constructing this array, the inventory has ten $\frac{1}{4}$-W 100-Ω resistors, since the number in column 1 ($\frac{1}{4}$ W) and row 1 (100 Ω) is 10. Similarly, column 3 (1 W) has an "8" in row 4 (33 kΩ), so there are eight 1-W 33-kΩ resistors.

A matrix of more than two dimensions can be constructed but cannot be easily represented on two-dimensional paper. The idea is simple, however; just add more rules to the construction. For example, let us assume that the resistor inventory is *further* distinguished by the accuracy rating 1%, 5%, and 10%. Now the inventory must tell the number of resistors with a certain resistance, certain power rating, and certain accuracy. This will require a third dimension to the matrix, which would be a 5 $\times$ 3 $\times$ 3 matrix. In network analysis it is generally sufficient to use one- and two-dimensional matrices.

From the definitions presented above, a general two-dimensional matrix of n rows and m columns ($n \times m$) can be written in the form

$$\mathbf{A} = \begin{pmatrix} a_{11} & a_{12} & a_{13} & \cdots & a_{1j} & \cdots & a_{1m} \\ a_{21} & a_{22} & a_{23} & \cdots & a_{2j} & \cdots & a_{2m} \\ a_{31} & a_{32} & \cdot & \cdots & a_{3j} & \cdots & a_{3m} \\ \cdot & \cdot & \cdot & \cdot & \cdot & \cdot & \cdot \\ \cdot & \cdot & \cdot & \cdot & \cdot & \cdot & \cdot \\ a_{i1} & a_{i2} & \cdot & \cdots & a_{ij} & \cdots & a_{im} \\ \cdot & \cdot & \cdot & \cdot & \cdot & \cdot & \cdot \\ \cdot & \cdot & \cdot & \cdot & \cdot & \cdot & \cdot \\ a_{n1} & a_{n2} & & \cdots & a_{nj} & \cdots & a_{nm} \end{pmatrix} \qquad (2\text{-}9)$$

The small a's are the *elements* of the matrix and are the actual numbers that make up the array. These elements are identified by two indices, as shown. The general element is a_{ij}. The first index, i, represents the row in which the element is found and the second index, j, represents the column in which the element is found. This matrix is of order $n \times m$. If $m = 1$, the result would be the column matrix introduced earlier. If $m = n$, the matrix is said to be *square* since it has as many rows as columns. Those elements that have $i = j$, such as a_{11} and a_{22}, are the *diagonal* elements of the matrix.

EXAMPLE 2-3

Given the matrix

$$\mathbf{B} = \begin{pmatrix} -5 & 0 & 6 & 1 \\ 4 & -3 & 0 & 3 \\ -2 & -1 & 2 & -6 \end{pmatrix}$$

identify the values of the elements b_{13}, b_{24}, and b_{32}.

SOLUTION

Using the definition given above for the general matrix, Eq. (2-9), b_{13} will be the number found in row 1 and column 3, which is a "6." Thus $b_{13} = 6$. By a similar process, $b_{24} = 3$ and $b_{32} = -1$. ∎

Products of Matrices. In the process of using arrays of numbers the concept of multiplication of these arrays was encountered. This came about in the context of having the array, or matrix, represent physical quantities such as weight, current, and speed. Even in the inventory example presented above, the *concept* of needing to multiply two matrices can be visualized. Suppose that the dollar value of the inventory is desired. If, in the two-dimensional case, the different resistance values and power ratings have different costs per unit, there should be way of multiplying a "cost per unit" matrix times the inventory matrix to obtain another matrix giving the dollar value of stock in each category. Indeed, such multiplication is possible.

One of the first results of studying how to multiply matrices is the discovery that a certain condition must be met before multiplication is possible:

Matrix Product Condition. *Multiplication of matrix* **A** *times matrix* **B** *is possible only when* **A** *has the same number of columns as* **B** *has rows.*

If this condition is not met, they cannot be multiplied; that is, the rules for multiplying the matrices do not apply. If this condition is satisfied, the product can be represented symbolically as

$$\mathbf{C} = \mathbf{AB} \tag{2-10}$$

where

$$\mathbf{A} = \text{matrix of order } n \times k$$

$$\mathbf{B} = \text{matrix of order } k \times m$$

$$\mathbf{C} = \text{matrix of order } n \times m$$

Notice that the product matrix, $\mathbf{C}$, has the same number of rows as $\mathbf{A}$ and the same number of columns as $\mathbf{B}$. The order of the product in matrix algebra is important. In Eq. (2-10) the reverse product (i.e., $\mathbf{BA}$) is not possible because the condition above would not be satisfied unless it was true that $n = m$.

The actual product is defined by a mathematical rule telling how to construct the elements of $\mathbf{C}$ from the elements of $\mathbf{A}$ and $\mathbf{B}$. Consider a general element of $\mathbf{C}$, given by c_{ij}. The value of this element is found by the sum of the products of the elements of row i of matrix $\mathbf{A}$ and the elements of column j of matrix $\mathbf{B}$. In equation form this is

$$c_{ij} = a_{i1}b_{1j} + a_{i2}b_{2j} + \cdots + a_{ik}b_{kj} \tag{2-11}$$

From this equation it is clear that if $\mathbf{A}$ does not have the same number of columns (k) as $\mathbf{B}$ has rows (k), then Eq. (2-11) will not work out since elements of $\mathbf{A}$ or $\mathbf{B}$ will get left out.

EXAMPLE 2-4

Given two matrices,

$$\mathbf{A} = \begin{pmatrix} 2 & 1 \\ 3 & 2 \\ 4 & 2 \end{pmatrix} \quad \text{and} \quad \mathbf{B} = \begin{pmatrix} 2 & -1 \\ 3 & 4 \end{pmatrix}$$

show that matrix multiplication is possible and find the product matrix, $\mathbf{C} = \mathbf{AB}$.

SOLUTION

To show that matrix multiplication is possible, the product must satisfy the condition given above. Since $\mathbf{A}$ has two columns and $\mathbf{B}$ has two rows, the condition is satisfied and multiplication is possible. We know also that $\mathbf{C}$ will have three rows and two columns. The product can be written

$$\begin{pmatrix} c_{11} & c_{12} \\ c_{21} & c_{22} \\ c_{31} & c_{32} \end{pmatrix} = \begin{pmatrix} 2 & 1 \\ 3 & 2 \\ 4 & 2 \end{pmatrix} \begin{pmatrix} 2 & -1 \\ 3 & 4 \end{pmatrix}$$

Now the definition of Eq. (2-11) can be used to find each element of $\mathbf{C}$.

$$c_{11} = a_{11}b_{11} + a_{12}b_{21} = 2(2) + 1(3) = 7$$

$$c_{12} = a_{11}b_{12} + a_{12}b_{22} = 2(-1) + 1(4) = 2$$

$$c_{21} = a_{21}b_{11} + a_{22}b_{21} = 3(2) + 2(3) = 12$$

$$c_{22} = a_{21}b_{12} + a_{22}b_{22} = 3(-1) + 2(4) = 5$$

$$c_{31} = a_{31}b_{11} + a_{32}b_{21} = 4(2) + 2(3) = 14$$

$$c_{32} = a_{31}b_{12} + a_{32}b_{22} = 4(-1) + 2(4) = 4$$

Thus the product matrix, **C**, can be written in the form

$$\mathbf{C} = \begin{pmatrix} 7 & 2 \\ 12 & 5 \\ 14 & 4 \end{pmatrix}$$

∎

Notice that if you try to multiply **B** times **A**, that is, in reverse order to that in the example, multiplication is not possible according to the rule of Eq. (2-11).

2-3.2 Simultaneous Equations

Much of network analysis is concerned with developing and then solving a set of simultaneous equations. This section shows how a set of simultaneous equations can be represented in matrix form and then reviews the Cramer's rule approach for finding solutions of the equations.

General Simultaneous Equations. The general form of a set of simultaneous equations can be written in terms of the unknowns, the coefficients of the unknowns, and the *forcing functions*, which are terms in the equations which do not involve the unknowns. A set of n equations in n unknowns has the form

$$
\begin{aligned}
r_{11}i_1 + r_{12}i_2 + r_{13}i_3 + \cdots + r_{1n}i_n &= v_1 \\
r_{21}i_1 + r_{22}i_2 + r_{23}i_3 + \cdots + r_{2n}i_n &= v_2 \\
r_{31}i_1 + r_{32}i_2 + \quad\cdots + r_{3n}i_n &= v_3 \\
&\vdots \\
r_{n1}i_1 + r_{n2}i_2 + \quad\cdots + r_{nn}i_n &= v_n
\end{aligned}
\tag{2-12}
$$

where

$$i_1, i_2, \ldots, i_n = \text{the } n \text{ unknowns}$$

$$v_1, v_2, \ldots, v_n = \text{the } n \text{ forcing functions}$$

$$r_{11}, r_{12}, \ldots, = \text{the } n^2 \text{ coefficients}$$

The general coefficient is labeled r_{ij}, where i denotes the equation in which the coefficient is found and j denotes which unknown the coefficient multiplies. The r_{ij} can be positive or negative numbers or zero, as can the forcing functions, v_i.

Matrix Representation. There is a close correspondence between a set of simultaneous equations and a matrix product equation. To see this, consider the product of a square, 3×3 matrix, **A,** with a column, 3×1 matrix, **B.** From the preceding section the result will be another column matrix, **C:**

$$C = AB$$

If this is written out in terms of the elements, the equation becomes

$$\begin{pmatrix} c_{11} \\ c_{21} \\ c_{31} \end{pmatrix} = \begin{pmatrix} a_{11} & a_{12} & a_{13} \\ a_{21} & a_{22} & a_{23} \\ a_{31} & a_{32} & a_{33} \end{pmatrix} \begin{pmatrix} b_{11} \\ b_{21} \\ b_{31} \end{pmatrix}$$

Equations can be written for the elements of the **C** matrix using the matrix multiplication rule given earlier. Doing this will give three equations, one for each element of **C:**

$$c_{11} = a_{11}b_{11} + a_{12}b_{21} + a_{13}b_{31}$$

$$c_{21} = a_{21}b_{11} + a_{22}b_{21} + a_{23}b_{31}$$

$$c_{31} = a_{31}b_{11} + a_{32}b_{21} + a_{33}b_{31}$$

When these equations are compared to the general set of simultaneous equations given above, it becomes clear that they have the *same form.* If the **B** matrix elements are defined to be the unknowns, the **A** matrix elements the coefficients, and the **C** matrix elements the forcing functions, the two have the same form.

With this description of the equivalence of representations it is easy to show that the simultaneous equations of Eq. (2-12) can be written in the following form as a matrix product:

$$\begin{pmatrix} r_{11} & r_{12} & r_{13} & \cdots & r_{1n} \\ r_{21} & r_{22} & & \cdots & r_{2n} \\ r_{31} & r_{32} & & \cdots & r_{3n} \\ \vdots & \vdots & \vdots & & \vdots \\ r_{n1} & r_{n2} & & \cdots & r_{nn} \end{pmatrix} \begin{pmatrix} i_1 \\ i_2 \\ i_3 \\ \vdots \\ i_n \end{pmatrix} = \begin{pmatrix} v_1 \\ v_2 \\ v_3 \\ \vdots \\ v_n \end{pmatrix} \qquad (2\text{-}13)$$

This matrix equation is the same as the set of simultaneous equations given in Eq. (2-12). This can easily be proven by multiplying out the matrix equation. The result will be the simultaneous equations of Eq. (2-12). This result is quite general since no restrictions have been placed on either the simultaneous set or the matrix equation, Eq. (2-13).

EXAMPLE 2-5

Given a set of simultaneous equations,

$$3x + 2y - 3z = 14$$
$$2x \quad\quad + 2z = 9$$
$$3x - 4y \quad\quad = -3$$

write the set of equations as a matrix equation.

SOLUTION

Referring to the general simultaneous equations of Eq. (2-12), it is clear that the coefficient matrix is just the coefficients of the unknowns in the equations; the unknown form a column matrix and the forcing functions the other column matrix. From inspection, then, the matrix equation is

$$\begin{pmatrix} 3 & 2 & -3 \\ 2 & 0 & 2 \\ 3 & -4 & 0 \end{pmatrix} \begin{pmatrix} x \\ y \\ z \end{pmatrix} = \begin{pmatrix} 14 \\ 9 \\ -3 \end{pmatrix}$$

■

EXAMPLE 2-6

Given the following matrix equation, find the set of simultaneous equations for a, b, and c.

$$\begin{pmatrix} 3 & -2 & 0 \\ 4 & 1 & 2 \\ 0 & 2 & 1 \end{pmatrix} \begin{pmatrix} a \\ b \\ c \end{pmatrix} = \begin{pmatrix} 6 \\ 0 \\ 4 \end{pmatrix}$$

SOLUTION

To find the simultaneous equations, it is only necessary to multiply the matrix equation out using the elements and the multiplication rule of matrices:

$$3a - 2b \quad\quad = 6$$
$$4a + b + 2c = 0$$
$$2b + c = 4$$

■

Cramer's Rule. The previous sections of this chapter showed how simultaneous equations could be solved by the method of substitution. Although this method will work for any number of simultaneous equations, it does become quite cumbersome when the number of equations exceeds three. Cramer's rule is a method of solving the equations which is easy to use when the equations have been written in a matrix representation. The calculations of Cramer's rule involve the evaluation of determinants, which will be reviewed in the next section.

Given a matrix equation in n unknowns such as Eq. (2-13), Cramer's rule defines the solution for any one of the unknowns, i_k as the quotient of two determinants. The form of the solution is

$$i_k = \frac{\begin{vmatrix} r_{11} & r_{12} & r_{13} & \cdots & v_1 & \cdots & r_{1n} \\ r_{21} & r_{22} & \cdots & \cdots & v_2 & \cdots & r_{2n} \\ r_{31} & r_{32} & \cdots & \cdots & v_3 & \cdots & r_{3n} \\ \cdot & \cdot & & & \cdot & & \cdot \\ \cdot & \cdot & & & \cdot & & \cdot \\ \cdot & \cdot & & & \cdot & & \cdot \\ r_{n1} & r_{n2} & \cdots & \cdots & v_n & \cdots & r_{nn} \end{vmatrix}}{\begin{vmatrix} r_{11} & r_{12} & r_{13} & \cdots & r_{1k} & \cdots & r_{1n} \\ r_{21} & r_{22} & & & r_{2k} & & r_{2n} \\ r_{31} & r_{32} & & & r_{3k} & & r_{3n} \\ \cdot & \cdot & & & \cdot & & \cdot \\ \cdot & \cdot & & & \cdot & & \cdot \\ \cdot & \cdot & & & \cdot & & \cdot \\ r_{n1} & r_{n2} & \cdots & \cdots & r_{nk} & \cdots & r_{nn} \end{vmatrix}} \tag{2-14}$$

The denominator is simply the determinant of the coefficient matrix. The numerator is the coefficient matrix, *but* the kth row, corresponding to the kth unknown i_k, has been replaced by the forcing function terms, v_1 to v_n. Therefore, the problem of finding the value of an unknown reduces to one of evaluation of two determinants. You have learned how to evaluate determinants, at least of order 2 and 3, in previous coursework. In general, a network problem may often result in determinants of higher order, so it will be important to know how to evaluate determinants of order greater than 3. The following problem illustrates Cramer's rule for solving a system of three unknowns. The next section reviews determinants and presents a technique for evaluation of higher-order determinants.

EXAMPLE 2-7

Find the solution of the following matrix equation using Cramer's rule.

$$\begin{pmatrix} 3 & 2 & -1 \\ 0 & -5 & 2 \\ 3 & 1 & 0 \end{pmatrix} \begin{pmatrix} x_1 \\ x_2 \\ x_3 \end{pmatrix} = \begin{pmatrix} 0 \\ 2 \\ 1 \end{pmatrix}$$

SOLUTION

Using the definition of Cramer's rule, the solutions for the unknowns can be found using Eq. (2-14).

$$x_1 = \frac{\begin{vmatrix} 0 & 2 & -1 \\ 2 & -5 & 2 \\ 1 & 1 & 0 \end{vmatrix}}{\begin{vmatrix} 3 & 2 & -1 \\ 0 & -5 & 2 \\ 3 & 1 & 0 \end{vmatrix}} = \frac{4 - 2 - 5}{12 - 15 - 6} = \frac{-3}{-9} = \frac{1}{3}$$

The denominator determinant turned out to be -9; this is in every term and will not have to be calculated again. The remaining unknowns are

$$x_2 = \frac{\begin{vmatrix} 3 & 0 & -1 \\ 0 & 2 & 2 \\ 3 & 1 & 0 \end{vmatrix}}{-9} = \frac{6 - 6}{-9} = 0$$

$$x_3 = \frac{\begin{vmatrix} 3 & 2 & 0 \\ 0 & -5 & 2 \\ 3 & 1 & 1 \end{vmatrix}}{-9} = \frac{-15 + 12 - 6}{-9} = 1$$

This example illustrates the steps involved in solving a matrix equation, or simultaneous set of equations, using Cramer's rule and determinants. In this case the determinants were 3×3 and quite easy to evaluate by hand. The following section reviews the concept of determinants and their evaluation for second, third, and higher orders.

2-3.3 Determinants

Determinants appear in many mathematical procedures employed to solve real-world problems. Generally speaking, they are most useful in problems involving systems of equations such as are found in network analysis.

Definition. A determinant is defined as a square array of numbers, $n \times n$, which has a specific single value representing the array. A general $n \times n$ determinant can be written in the form

$$D = \begin{vmatrix} r_{11} & r_{12} & r_{13} & \cdots & & r_{1n} \\ r_{21} & r_{22} & \cdot & \cdots & & r_{2n} \\ r_{31} & r_{32} & \cdot & \cdots & & r_{3n} \\ \cdot & \cdot & \cdot & & & \cdot \\ \cdot & \cdot & \cdot & & & \cdot \\ \cdot & \cdot & \cdot & & & \cdot \\ r_{n1} & r_{n2} & \cdot & \cdots & & r_{nn} \end{vmatrix} \qquad (2\text{-}15)$$

The straight lines distinguish a determinant from a matrix. Remember that a matrix is an array of numbers but it does not have a specific value. Notice that a determinant is a *square* array of numbers because it has as many rows as columns. The *order* of a determinants is the number of rows or columns, so the determinant above is of order n. A determinant has a *value* deduced from the elements of the array by specific sums and differences of products of the elements. The formal definition tells what type of products to make and whether to add or subtract, but provides no organized method for constructing the required terms. Fortunately, several methods are available to do this.

2 × 2 and 3 × 3 Determinants. It is a matter of mathematical coincidence that 2 × 2 and 3 × 3 determinants can be evaluated by following simple sum and difference rules for diagonal products of elements. You must remember that this method will work *only* for these two determinants.

For the 2 × 2 determinant, the solution is simply the difference of products of the right and left diagonal elements:

$$D = \begin{vmatrix} a & b \\ c & d \end{vmatrix} = ad - bc \qquad (2\text{-}16)$$

EXAMPLE 2-8

Evaluate the determinant

$$\begin{vmatrix} 6 & 3 \\ 2 & 4 \end{vmatrix}$$

SOLUTION

From Eq. (2-16) the solution is

$$\begin{vmatrix} 6 & 3 \\ 2 & 4 \end{vmatrix} = 6(4) - 3(2) = 18$$

■

For 3 × 3 determinants evaluation is accomplished by copying the first and second columns after the last column and then taking sums and differences of diagonal products:

$$
D = \begin{vmatrix} a & d & g \\ b & e & h \\ c & f & i \end{vmatrix} \begin{matrix} a & d \\ b & e \\ c & f \end{matrix}
$$

$$
= (aei + dhc + gbf) - (gec + ahf + dbi)
$$

(2-17)

EXAMPLE 2-9

Evaluate

$$
\begin{vmatrix} 3 & 3 & 1 \\ 2 & -1 & 2 \\ 4 & 1 & 0 \end{vmatrix}
$$

SOLUTION

Using Eq. (2-17) as a guide, the evaluation is

$$
D = \begin{vmatrix} 3 & 3 & 1 \\ 2 & -1 & 2 \\ 4 & 1 & 0 \end{vmatrix} \begin{matrix} 3 & 3 \\ 2 & -1 \\ 4 & 1 \end{matrix}
$$

$$
= 3(-1)(0) + 3(2)(4) + 1(2)(1) - [1(-1)(4)
$$

$$
+ 3(2)(1) + 3(2)(0)]
$$

Therefore, the solution is $D = 26 - 2 = 24$. ∎

It is important to note that the foregoing two simple rules for determinant evaluation *work only* for determinants of order 2 and 3. Higher-order determinants must be evaluated by other means such as the expansion technique presented next.

Evaluation of Higher-Order Determinants. The evaluation of higher-order determinants must proceed according to the definition of the meaning of a determinant. No simple technique exists. However, to simplify the process of accounting for all the necessary terms, a number of formal approaches have been developed. The one that is perhaps easiest for hand evaluation is called the *Laplace expansion*. This method is sometimes referred to as *order reduction* since it expresses a determinant of order n by a sum of n determinants of order $n - 1$. Thus a 4 × 4 determinant would become four 3 × 3 determinants. We can then evaluate the third-order determinants using the technique shown above.

The first step in Laplace expansion is to define a number called the *cofactor* of an element of a determinant. Each element r_{ij} of the determinant has a co-factor, R_{ij}, defined as follows:

$$
R_{ij} = (-1)^{i+j}
\begin{vmatrix}
r_{11} & r_{12} & \cdots & r_{1,j-1} & r_{1,j+1} & \cdots & r_{1n} \\
r_{21} & r_{22} & \cdots & r_{2,j-1} & r_{2,j+1} & \cdots & r_{2n} \\
r_{31} & & \cdot & \cdot & \cdot & \cdot & \cdot \\
\cdot & & & & & & \cdot \\
\cdot & \cdot & \cdot & & & \cdot & \\
r_{i-1,1} & r_{i-1,2} & \cdot & & & \cdot & r_{i-1,n} \\
r_{i+1,1} & r_{i+1,2} & & & & \cdot & r_{i+1,n} \\
\cdot & & & & & & \\
\cdot & & & & & & \\
r_{n1} & r_{n2} & \cdot & r_{n,j-1} & r_{n,j+1} & \cdot & r_{nn}
\end{vmatrix}
\qquad (2\text{-}18)
$$

The first part of this expression, $(-1)^{i+j}$, determines only a sign (i.e., $+1$ or -1) to be multiplied times the determinant that follows. The determinant itself in Eq. (2-18) is seen to be the original determinant, Eq. (2-15), but with row i and column j taken out. Since a row and a column of the original determinant have been taken out, the result is a determinant of order one less than the original. The following example illustrates how the cofactors are formed. Following this it will be shown how Laplace expansion uses the cofactors to evaluate a determinant.

EXAMPLE 2-10

For the following 4×4 determinant, find the cofactors R_{11}, R_{32}, and R_{24}.

$$
\begin{vmatrix}
3 & 1 & 2 & 0 \\
-1 & 4 & 0 & 2 \\
1 & 1 & 3 & 0 \\
4 & -3 & 2 & 1
\end{vmatrix}
$$

SOLUTION

The definition of cofactor is given in Eq. (2-18). For the R_{11} term the result is

$$
R_{11} = (-1)^{1+1}
\begin{vmatrix}
4 & 0 & 2 \\
1 & 3 & 0 \\
-3 & 2 & 1
\end{vmatrix}
$$

Notice that the elements of row 1 and column 1 are missing from this determinant and consequently the determinant has an order of 3. It can be evaluated by the simple rule

$$R_{11} = (-1)^2[12 + 4 - (-18)] = (+1)(16 + 18)$$

$$= 34$$

To find R_{32} it will be necessary to take row 3 and column 2 from the original determinant. This leaves the expression

$$R_{32} = (-1)^{3+2} \begin{vmatrix} 3 & 2 & 0 \\ -1 & 0 & 2 \\ 4 & 2 & 1 \end{vmatrix}$$

$$= (-1)^5[16 - (-2 + 12)] = (-1)(16 - 10)$$

$$= -6$$

For R_{24}, row 2 and column 4 are taken out, leaving

$$R_{24} = (-1)^{2+4} \begin{vmatrix} 3 & 1 & 2 \\ 1 & 1 & 3 \\ 4 & -3 & 2 \end{vmatrix}$$

$$= (-1)^6[6 + 12 - 6 - (8 - 27 + 2)]$$

$$= (+1)(12 + 17) = 29 \qquad \blacksquare$$

Laplace Expansion.   The method of Laplace expansion evaluates a determinant using the cofactors, as defined above. The formal definition is:

Laplace Expansion.   The value of a determinant is given by the sum of elements times their respective cofactors, taken along all the elements in a single row or a single column.

So in this procedure, you select any row or any column. Then for each element of the row or column selected, the cofactor is calculated, as above. The value of the determinant is then the sum of the elements times the cofactors. In equation form this law is written as follows, first assuming, along row i,

$$D = r_{i1}R_{i1} + r_{i2}R_{i2} + \cdots + r_{in}R_{in} \qquad (2\text{-}19)$$

For evaluation along a column, j, the equation is

$$D = r_{1j}R_{1j} + r_{2j}R_{2j} + \cdots + r_{nj}R_{nj} \qquad (2\text{-}20)$$

It makes no difference which row or column is chosen for the expansion. It can make a difference in terms of how much calculation is necessary, as illustrated by the following example.

EXAMPLE 2-11

Evaluate the determinant

$$\begin{vmatrix} 3 & -1 & 2 & 4 \\ 4 & 0 & -2 & 4 \\ 0 & 3 & 0 & 2 \\ 1 & 4 & 0 & 2 \end{vmatrix}$$

SOLUTION

According to Laplace expansion the value of this determinant can be found by a sum of elements times cofactors along any row or column. Suppose that row 1 is chosen. Using Eq. (2-19), the value is

$$D = r_{11}R_{11} + r_{12}R_{12} + r_{13}R_{13} + r_{14}R_{14}$$

First the elements will be substituted:

$$D = 3R_{11} - R_{12} + 2R_{13} + 4R_{14}$$

Now it will be necessary to evaluate the cofactors, which are each 3×3 determinants. But wait! Since *any* row or column can be used, suppose that row 3 is selected. Then the determinant is

$$D = r_{31}R_{31} + r_{32}R_{32} + r_{33}R_{33} + r_{34}R_{34}$$

Again, substituting the values of the elements yields

$$D = (0)R_{31} + 3R_{32} + (0)R_{33} + 2R_{34}$$

$$= 3R_{32} + 2R_{34}$$

Notice that the evaluation of the determinant is reduced to finding the value of two cofactors only, which means two 3×3 determinants instead of four. From this you can see that the row or column should be selected which has the *most zeros* as elements.

Now, finding the cofactors,

$$R_{32} = (-1)^{3+2} \begin{vmatrix} 3 & 2 & 4 \\ 4 & -2 & 4 \\ 1 & 0 & 2 \end{vmatrix} = 12$$

and

$$R_{34} = (-1)^{3+4} \begin{vmatrix} 3 & -1 & 2 \\ 4 & 0 & -2 \\ 1 & 4 & 0 \end{vmatrix} = -58$$

So the solution becomes

$$D = 3(12) + 2(-58)$$
$$= -80$$

To illustrate that any row or column can be used, a column will now be selected and the evaluation repeated. Which column should be selected? Column 3, of course, since it has two zeros and will be the easiest to use. From Eq. (2-20),

$$D = r_{13}R_{13} + r_{23}R_{23} + r_{33}R_{33} + r_{43}R_{43}$$
$$= 2R_{13} - 2R_{23}$$

For the cofactors,

$$R_{13} = (-1)^{1+3} \begin{vmatrix} 4 & 0 & 4 \\ 0 & 3 & 2 \\ 1 & 4 & 2 \end{vmatrix} = -20$$

$$R_{23} = (-1)^{2+3} \begin{vmatrix} 3 & -1 & 4 \\ 0 & 3 & 2 \\ 1 & 4 & 2 \end{vmatrix} = 20$$

Thus the determinant is

$$D = 2(-20) - 2(20) = -80 \qquad \blacksquare$$

As this example illustrates, Laplace expansion provides a procedure by which a determinant can be evaluated. The example also demonstrates the value of having zeros as elements in the determinant and of using the row or column containing the most zeros for expansion. This becomes particularly important when considering determinants of high order. Consider for a moment, that a determinant of order 6 would be expanded to six determinants of order 5. Each of these would be expanded to five determinants of order 4 and each of these to four determinants of order 3. Whew! We would end up with 120 3 × 3 determinants to evaluate. However, for every zero in a row or column of the original along which expansion is performed, 20 3 × 3 determinants would be eliminated! You can see the value of zeros. In the next section a technique is presented by which zeros can be *generated* in a determinant.

Zero Generation. A special property of determinants makes the application of Laplace expansion particularly attractive. This property is that it is possible to generate zeros in elements of a determinant in such a way that the value of the determinant is not changed. The generation is accomplished by using the following general property of determinants:

Determinant Property. *A multiple of the elements of any row (or column) can be added to or subtracted from the elements of another row (or column) without changing the value of the determinant.*

Note that the row or column statement is exclusive, that is, row to row or column to column but not mixed. A multiple means any rational or irrational number multiplied times the elements of the row or column selected. The resulting numbers are added or subtracted, element by element, with the row or column selected to be operated upon. The following examples illustrate this process.

EXAMPLE 2-12

Evaluate the following determinant using zero generation and Laplace expansion.

$$\begin{vmatrix} 3 & 2 & -4 & 4 \\ 2 & -1 & 2 & 4 \\ 3 & -1 & -1 & 2 \\ 1 & 4 & 1 & 2 \end{vmatrix}$$

SOLUTION

The first problem will be to generate zeros in the determinant. The objective is to get a row or column with all but one element zero, since then the value of the determinant is found by evaluation of one cofactor. Notice that if row 3 is simply added to row 4, a zero is generated in column 3. Note that the row or column being used is not changed by this operation.

$$\begin{vmatrix} 3 & 2 & -4 & 4 \\ 2 & -1 & 4 & 4 \\ 3 & -1 & -1 & 2 \\ 1+3 & 4-1 & 1-1 & 2+2 \end{vmatrix}$$

or

$$\begin{vmatrix} 3 & 2 & -4 & 4 \\ 2 & -1 & 2 & 4 \\ 3 & -1 & -1 & 2 \\ 4 & 3 & 0 & 4 \end{vmatrix}$$

Notice that a zero has been generated in column 3 at row 4, but the value of the determinant is unchanged. Now a multiple of two times row 3 will be added to row 2. This selection is made because you can see that this will generate a zero in column 3 at row 2:

$$\begin{vmatrix} 3 & 2 & -4 & 4 \\ 2 + 2(3) & -1 + 2(-1) & 2 + 2(-1) & 4 + 2(2) \\ 3 & -1 & -1 & 2 \\ 4 & 3 & 0 & 4 \end{vmatrix}$$

or

$$\begin{vmatrix} 3 & 2 & -4 & 4 \\ 8 & -3 & 0 & 8 \\ 3 & -1 & -1 & 2 \\ 4 & 3 & 0 & 4 \end{vmatrix}$$

So now column 3 has another zero. It is clear that another zero can be generated in this column by subtracting four times row 3 from row 1:

$$\begin{vmatrix} 3 - 4(3) & 2 - 4(-1) & -4 - 4(-1) & 4 - 4(2) \\ 8 & -3 & 0 & 8 \\ 3 & -1 & -1 & 2 \\ 4 & 3 & 0 & 4 \end{vmatrix}$$

or

$$\begin{vmatrix} -9 & 6 & 0 & -4 \\ 8 & -3 & 0 & 8 \\ 3 & -1 & -1 & 2 \\ 4 & 3 & 0 & 4 \end{vmatrix}$$

Of course, since column 3 now has only one nonzero element, a Laplace expansion will be done on this column, giving the value of the determinant as

$$D = r_{33}R_{33} = (-1)R_{33} = -R_{33}$$

The cofactor R_{33} is found from Eq. (2-18):

$$R_{33} = (-1)^{3+3} \begin{vmatrix} -9 & 6 & -4 \\ 8 & -3 & 8 \\ 4 & 3 & 4 \end{vmatrix}$$

This 3 × 3 determinant can simply be multiplied out, but notice that if column 1 is subtracted from column 3, two zeros are generated:

$$R_{33} = \begin{vmatrix} -9 & 6 & 5 \\ 8 & -3 & 0 \\ 4 & 3 & 0 \end{vmatrix} = 120 + 60 = 180$$

The value of the determinant is now

$$D = (-1)(180) = -180 \qquad \blacksquare$$

This example shows that zero generation combined with Laplace expansion form a powerful combination for evaluation of higher-order determinants. In the latter case a 4 × 4 was finally evaluated as a single 3 × 3 determinant.

EXAMPLE 2-13

Evaluate the determinant

$$\begin{vmatrix} 3 & 0 & 0 & 2 & 1 \\ 4 & 0 & -3 & 6 & 0 \\ -2 & 6 & 8 & 4 & -1 \\ 0 & 2 & 0 & 4 & 2 \\ 1 & 0 & 8 & -2 & 0 \end{vmatrix}$$

SOLUTION

Column 2 already has three zeros, so an obvious choice is to generate another zero. This can be done by subtracting three times row 4 from row 3.

$$\begin{vmatrix} 3 & 0 & 0 & 2 & 1 \\ 4 & 0 & -3 & 6 & 0 \\ -2 & 0 & 8 & -8 & -7 \\ 0 & 2 & 0 & 4 & 2 \\ 1 & 0 & 8 & -2 & 0 \end{vmatrix}$$

Laplace expansion on column 2 now gives

$$D = r_{42}R_{42} = 2R_{42}$$

$$R_{42} = (-1)^{4+2} = \begin{vmatrix} 3 & 0 & 2 & 1 \\ 4 & -3 & 6 & 0 \\ -2 & 8 & -8 & -7 \\ 1 & 8 & -2 & 0 \end{vmatrix}$$

To evaluate this 4×4 determinant, a zero will be generated in column 4 by adding seven times row 1 to row 3. The result is

$$\begin{vmatrix} 3 & 0 & 2 & 1 \\ 4 & -3 & 6 & 0 \\ 19 & 8 & 6 & 0 \\ 1 & 8 & -2 & 0 \end{vmatrix}$$

A Laplace expansion on column 4 gives the final result as a third-order determinant:

$$D = (2)(-1)^{4+2}(1)(-1)^{4+1} \begin{vmatrix} 4 & -3 & 6 \\ 19 & 8 & 6 \\ 1 & 8 & -2 \end{vmatrix}$$

$$= -952 \qquad \blacksquare$$

Note that in this example a 5×5 determinant was reduced to a single 3×3 determinant.

2-4

MESH AND NODAL ANALYSIS

When the two network laws presented earlier, KVL and KCL, are combined with the matrix representation of simultaneous equations, two very powerful approaches to network analysis can be developed. Each is based on observations about the nature of the equations that result from applying KVL and KCL to networks. Certain regular results are noted. These are formulated into a set of rules that can be used to construct the simultaneous equations, represented in matrix format. Once these rules are learned, it is usually possible to write the matrix equation out by inspection of the network, with little or no intermediate calculations.

2-4.1 Mesh Analysis

Mesh analysis results from observations made about the nature of equations developed from networks where KVL was used to generate the equations. Before giving the formal definition of mesh analysis it is of some value to see how these observations appear. This can be done by considering the equations that resulted from applying KVL to the network of Fig. 2–1. Remember that i_1 and i_2 turned out to be the required loop currents to solve the network. When Eqs. (2-4) and (2-5), which resulted from loop analysis of the network, are written in the matrix representation, the result is

$$\begin{pmatrix} R_1 + R_3 + R_4 & -R_3 \\ -R_3 & R_2 + R_3 + R_5 \end{pmatrix} \begin{pmatrix} i_1 \\ i_2 \end{pmatrix} = \begin{pmatrix} v_1 \\ -v_2 \end{pmatrix} \qquad (2\text{-}21)$$

Note that the diagonals of the coefficient matrix are the positive sum of resistors which are in the elementary loops corresponding in number to the diagonal. That is, the $(1, 1)$ element is the positive sum of resistors in loop 1. The off-diagonal elements are the negative of the resistance shared by the two loops. Thus both i_1 and i_2 go through R_3, and this resistor shows up as a negative in the $(1, 2)$ position of the matrix. From observations such as this, taken over many examples, and careful study of the circuit laws, a general pattern emerges on how the elements of the matrices are related to the network elements and the elementary loop currents. Using these patterns, rules are developed allowing easy determination of the coefficient and forcing function matrices. An important restriction on this method is that the network can contain only voltage sources. In order to construct the mesh matrix equation using the rules below any current sources *must* be transformed to voltage sources.

Rules for Mesh Analysis.   Given a resistive network of n elementary loops which contains *only* voltage sources:

1. Identify each elementary loop and assign to each a *clockwise* loop current, numbered 1 through n in any order. These loops form the "mesh" of the network and the loop currents are the mesh currents.
2. The matrix equation for the unknown currents will have the form

$$\begin{pmatrix} r_{11} & r_{12} & \cdots & r_{1n} \\ r_{21} & r_{22} & \cdots & r_{2n} \\ r_{31} & & & \\ \vdots & \vdots & & \vdots \\ r_{n1} & r_{n2} & \cdots & r_{nn} \end{pmatrix} \begin{pmatrix} i_1 \\ i_2 \\ \vdots \\ i_n \end{pmatrix} = \begin{pmatrix} v_1 \\ v_2 \\ \vdots \\ v_n \end{pmatrix} \qquad (2\text{-}22)$$

3. $i_1, i_2, \ldots, i_n$ = unknown elementary loop currents.
4. r_{ii} = diagonal elements = positive sum of the resistors in loop i.
5. r_{ij} = off-diagonal elements = negative of resistors shared by loops i and j. Note that $r_{ij} = r_{ji}$. This means that the coefficient matrix is symmetric about the diagonal.
6. v_i = forcing functions = sum of sources in loop i. The polarity is determined such that if the loop current leaves the positive terminal of the source, it is added to the term, and if the loop current enters the positive terminal of the source, it is subtracted from the term.

The following example illustrates in detail the application of mesh analysis to a network problem.

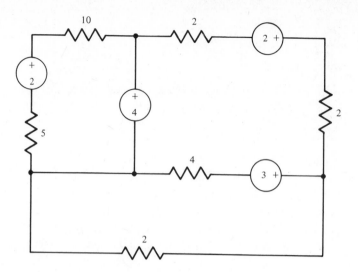

FIGURE 2–5. Network for Example 2-14.

EXAMPLE 2-14

Find the solution of the network in Fig. 2–5 using mesh analysis.

SOLUTION

In Fig. 2–6 the network has been redrawn with each elementary loop identified and assigned a clockwise loop current as specified in step 1 of the mesh rules. Note that there are three loops and three currents. Thus the matrix equation will have the form

$$\begin{pmatrix} r_{11} & r_{12} & r_{13} \\ r_{21} & r_{22} & r_{23} \\ r_{31} & r_{32} & r_{33} \end{pmatrix} \begin{pmatrix} i_1 \\ i_2 \\ i_3 \end{pmatrix} = \begin{pmatrix} v_1 \\ v_2 \\ v_3 \end{pmatrix}$$

The values of the elements of the coefficient matrix and the forcing function matrix are found using the remaining steps of the rules. For the diagonals,

$$r_{11} = \text{sum of resistors in loop 1}$$

$$= 5 + 10 = 15$$

$$r_{22} = \text{sum of resistors in loop 2}$$

$$= 2 + 2 + 4 = 8$$

$$r_{33} = \text{sum of resistors in loop 3}$$

$$= 4 + 2 = 6$$

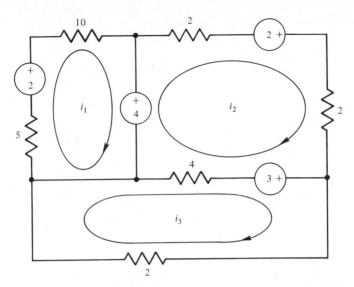

FIGURE 2–6. Network of Fig. 2–5 with loop currents assigned for mesh analysis.

For the off-diagonals,

$$r_{12} = r_{21} = \text{negative of resistors shared by loops 1 and 2}$$

$$= 0 \qquad \text{since the loops share a voltage source which has a zero internal resistance}$$

$$r_{13} = r_{31} = \text{negative of resistors shared by loops 1 and 3}$$

$$= 0 \qquad \text{since the loops do not share resistors}$$

$$r_{23} = r_{32} = \text{negative of resistors shared by loops 3 and 2}$$

$$= -4 \qquad \text{since they share a 4-}\Omega\text{ resistor}$$

For the forcing function matrix,

$$v_1 = +2 - 4 = -2$$
> The loop current, i_1, leaves the positive terminal of the 2-V source, so it is added; but it enters the positive terminal of the 4-V source, so it is subtracted.

$$v_2 = +4 + 2 - 3 = 3$$
$$v_3 = \quad 3$$

With the matrix elements identified, the matrix equation can now be written:

$$\begin{pmatrix} 15 & 0 & 0 \\ 0 & 8 & -4 \\ 0 & -4 & 6 \end{pmatrix} \begin{pmatrix} i_1 \\ i_2 \\ i_3 \end{pmatrix} = \begin{pmatrix} -2 \\ 3 \\ 3 \end{pmatrix}$$

Solution for the currents follows from Cramer's rule:

$$i_1 = \frac{\begin{vmatrix} -2 & 0 & 0 \\ 3 & 8 & -4 \\ 3 & -4 & 6 \end{vmatrix}}{\begin{vmatrix} 15 & 0 & 0 \\ 0 & 8 & -4 \\ 0 & -4 & 6 \end{vmatrix}} = \frac{-96 + 32}{480} = -0.13 \text{ A}$$

$$i_2 = \frac{\begin{vmatrix} 15 & -2 & 0 \\ 0 & 3 & -4 \\ 0 & 3 & 6 \end{vmatrix}}{480} = \frac{270 + 180}{480} = 0.94 \text{ A}$$

$$i_3 = \frac{\begin{vmatrix} 15 & 0 & -2 \\ 0 & 8 & 3 \\ 0 & -4 & 3 \end{vmatrix}}{480} = \frac{360 + 180}{480} = 1.13 \text{ A}$$ ∎

This example had only constant sources. However, mesh analysis works equally well for time-dependent sources and controlled sources. In some cases current sources will be found in a network for which mesh analysis is to be used. In these cases the current sources must be transformed into voltage sources. It is very important to remember that such a transformation has been made and to draw conclusions about electrical behavior within the transformed part of the network in terms of the original elements. This is illustrated by the following example.

EXAMPLE 2-15

Given the network of Fig. 2–7, find the current through the 3-Ω resistor and the output voltage, v_{out}.

SOLUTION

First, the 2-A current source must be transformed into a voltage source. This can be easily done since the source is shunted by a 3-Ω resistor. It will become a 6-V source in series with the 3-Ω resistor, as shown in Fig. 2–8, together with the assigned clockwise currents. A dashed circle has been drawn around

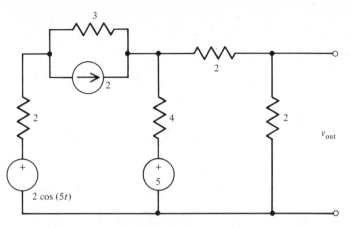

FIGURE 2–7.   Network for Example 2-15. This network contains a current source.

the transformed part of the network as a reminder that this part has been transformed. Now by following the mesh rules you can prove that the following matrix equation results:

$$\begin{pmatrix} 9 & -4 \\ -4 & 8 \end{pmatrix}\begin{pmatrix} i_1 \\ i_2 \end{pmatrix} = \begin{pmatrix} 2\cos(5t) + 1 \\ 5 \end{pmatrix}$$

To find the current through the 3-Ω resistor it would be natural to conclude that it was i_1. But this is not true! The reason is that the 3-Ω resistor is in the *transformed* part of the network. To find this current the solution for i_1 must be

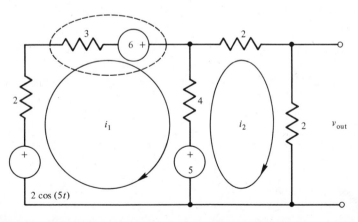

FIGURE 2–8.   The current source of the network in Fig. 2–7 has been transformed into a voltage source and mesh currents have been assigned.

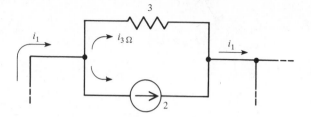

FIGURE 2–9. The original circuit must be used to find the current through the 3-Ω resistor.

used in the *original* network. Figure 2–9 shows the appropriate part of the original network, from Fig. 2–7, and the current i_1. From this, and KCL applied to the junction, you can clearly see that the required current is

$$i_{3\Omega} = i_1 - 2$$

Solving for i_1 with Cramer's rule gives

$$i_1 = \frac{\begin{vmatrix} 2\cos(5t) + 1 & -4 \\ 5 & 8 \end{vmatrix}}{\begin{vmatrix} 9 & -4 \\ -4 & 8 \end{vmatrix}}$$

$$= \frac{16\cos(5t) + 8 + 20}{72 - 16}$$

$$= 0.29\cos(5t) + 0.5 \text{ A}$$

So the current through the 3-Ω resistor becomes

$$i_{3\Omega} = 0.29\cos(5t) - 1.5 \text{ A}$$

Now the output voltage is found from Ohm's law, as $v_{out} = 2i_2$:

$$i_2 = \frac{\begin{vmatrix} 9 & 2\cos(5t) + 1 \\ -4 & 5 \end{vmatrix}}{56} = 0.14\cos(5t) + 0.875$$

Therefore, the output voltage is

$$v_{out} = 0.28\cos(5t) + 1.75 \text{ V}$$

2-4.2 Nodal Analysis

In the first part of this chapter, KCL was used to generate a set of simultaneous equations for the network of Fig. 2–4. The resulting set of equations can be written in matrix representation as

$$
\begin{pmatrix}
\dfrac{1}{R_1} & -\dfrac{1}{R_1} & 0 \\[2ex]
-\dfrac{1}{R_1} & \dfrac{1}{R_1} + \dfrac{1}{R_2} + \dfrac{1}{R_3} & -\dfrac{1}{R_2} \\[2ex]
0 & -\dfrac{1}{R_2} & \dfrac{1}{R_2}
\end{pmatrix}
\begin{pmatrix}
v_1 \\[2ex] v_2 \\[2ex] v_3
\end{pmatrix}
=
\begin{pmatrix}
i_1 \\[2ex] 0 \\[2ex] -i_2
\end{pmatrix}
\qquad (2\text{-}23)
$$

Just as in the case of application of KVL, the results of application of KCL to networks results in equations with certain obvious regularity. This is exploited to construct a set of rules by which the matrix equation can be constructed by inspection from a network. The first thing to notice from Eq. (2-33) is that the resistances all appear as their inverses, the conductance. Therefore, when using nodal analysis it will be helpful to express all resistance R as the conductance, $G = 1/R$. Notice that the diagonal terms in Eq. (2-23) are positive sums of conductances and the off-diagonal elements are negatives of conductances. The forcing function matrix consists of current sources. From these observations the following set of rules have emerged for construction of the matrix equation that will solve a network. In order to apply the following rules exactly as stated, the network must contain *only* current sources. If any voltage sources are present, they must be converted to current sources.

Rules for Nodal Analysis. Given a network of $n + 1$ nodes which contains *only* current sources:

1. Select one node as a reference. Assign n unknown voltages to the other nodes, in any order, as $v_1, v_2, \ldots, v_n$.
2. The matrix equation will be of the form

$$
\begin{pmatrix}
g_{11} & g_{12} & g_{13} & \cdots & g_{1n} \\
g_{21} & g_{22} & & \cdots & \\
g_{31} & & & \cdots & \\
 & & & \cdots & \\
 & & & \cdots & \\
 & & & \cdots & \\
g_{n1} & & & \cdots & g_{nn}
\end{pmatrix}
\begin{pmatrix}
v_1 \\ v_2 \\ \cdot \\ \cdot \\ \cdot \\ \cdot \\ v_n
\end{pmatrix}
=
\begin{pmatrix}
i_1 \\ i_2 \\ \cdot \\ \cdot \\ \cdot \\ \cdot \\ i_n
\end{pmatrix}
\qquad (2\text{-}24)
$$

3. $v_1, v_2, \ldots, v_n$ = unknown node voltages.
4. g_{ii} = diagonal elements = positive sum of the conductances directly connected to node i.
5. g_{ij} = off-diagonal elements = negative of conductances directly connecting node i to node j. Note that $g_{ij} = g_{ji}$, so that the coefficient matrix is symmetric about the diagonal.
6. i_i = forcing function = sum of current sources directly connected to node i. If the current source is directed into the node, it is added to the term, and if the source is directed out of the node, it is subtracted from the term.

When applying nodal analysis to a network it is important that all sources be current sources. The following examples illustrate application of this method to solving a network.

EXAMPLE 2-16
Find the solution of the network given in Fig. 2–10 using nodal analysis.

SOLUTION
In Fig. 2–11 the network has been redrawn with the nodes identified, the reference selected, and the voltage source transformed into a current source. To make the matrix construction easier, the resistances have been replaced by the conductance values. Note that the choice of a reference node is *arbitrary*. Whichever node makes the analysis easier can be chosen as the reference. The matrix equation will have the form

$$\begin{pmatrix} g_{11} & g_{12} & g_{13} \\ g_{21} & g_{22} & g_{23} \\ g_{31} & g_{32} & g_{33} \end{pmatrix} \begin{pmatrix} v_1 \\ v_2 \\ v_3 \end{pmatrix} = \begin{pmatrix} i_1 \\ i_2 \\ i_3 \end{pmatrix}$$

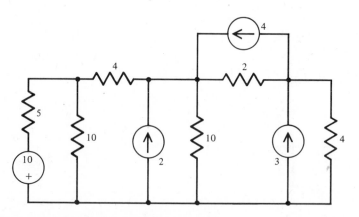

FIGURE 2–10. Network for Example 2-16.

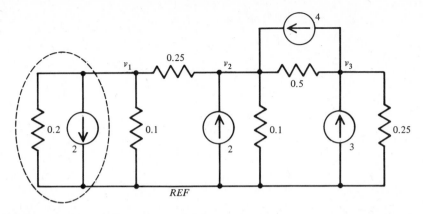

FIGURE 2–11. The voltage source in the network of Fig. 2–10 has been transformed into a current source and the reference node has been identified. Node voltages have been assigned for nodal analysis.

The values of the coefficient matrix and forcing function matrix elements are found from the rules above for nodal analysis.

g_{11} = sum of conductances connected to node 1

$= 0.2 + 0.1 + 0.25 = 0.55$

g_{22} = sum of conductances connected to node 2

$= 0.25 + 0.1 + 0.5 = 0.85$

g_{33} = sum of conductances connected to node 3

$= 0.5 + 0.25 = 0.75$

$g_{12} = g_{21}$ = negative of conductances joining nodes 1 and 2; they are joined by a 0.25 Sieman (S) conductance

$= -0.25$

$g_{13} = g_{31}$ = negative of conductances joining nodes 1 and 3; they are not directly joined

$= 0$

$g_{23} = g_{32}$ = negative of conductances joining nodes 2 and 3; they are joined by a 0.5 S conductance

$= -0.5$

The forcing functions are found by

i_1 = sum of current sources connected to node 1

$= -2$ since the 2-A source is directed out of the node

i_2 = sum of current sources connected to node 2

$\quad$ = $+2 + 4 = 6$ $\quad$ since both sources are directed into node 2

i_3 = sum of current sources connected to node 3

$\quad$ = $-4 + 3 = -1$ $\quad$ since the 4-A source is directed out and
$\qquad\qquad\qquad\qquad\qquad$ the 3-A source is directed into the node

Now the matrix equation can be written:

$$\begin{pmatrix} 0.55 & -0.25 & 0 \\ -0.25 & 0.85 & -0.5 \\ 0 & -0.5 & 0.75 \end{pmatrix} \begin{pmatrix} v_1 \\ v_2 \\ v_3 \end{pmatrix} = \begin{pmatrix} -2 \\ 6 \\ -1 \end{pmatrix}$$

Cramer's rule now allows the solutions to be found.

$$v_1 = \frac{\begin{vmatrix} -2 & -0.25 & 0 \\ 6 & 0.85 & -0.5 \\ -1 & -0.5 & 0.75 \end{vmatrix}}{\begin{vmatrix} 0.55 & -0.25 & 0 \\ -0.25 & 0.85 & -0.5 \\ 0 & -0.5 & 0.75 \end{vmatrix}} = \frac{0.225}{0.166} = 1.36$$

$$v_2 = \frac{\begin{vmatrix} 0.55 & -2 & 0 \\ -0.25 & 6 & -0.5 \\ 0 & -1 & 0.75 \end{vmatrix}}{0.166} = \frac{1.825}{0.166} = 10.99$$

$$v_3 = \frac{\begin{vmatrix} 0.55 & -0.25 & -2 \\ -0.25 & 0.85 & 6 \\ 0 & -0.5 & -1 \end{vmatrix}}{0.166} = \frac{0.995}{0.166} = 5.99$$ ∎

2-4.3 Controlled Sources

The presence of controlled sources does not alter the methods of mesh and nodal analysis for construction of the matrix equations. However, it is found that an *adjustment* of the matrix is required after it is constructed before the problem can be solved.

Consider the network of Fig. 2–12. Note the ICVS whose value of voltage is given by 12 times the net current through the 4-Ω resistor. The loop currents for mesh analysis have already been identified and labeled. Now the matrix equation is formed in the usual way, treating the controlled source like any other.

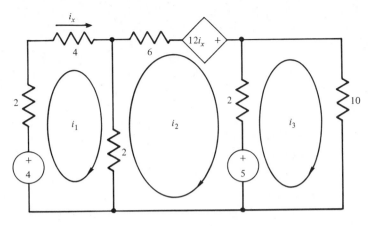

FIGURE 2–12. This network has a controlled source. Loop currents have been assigned for mesh analysis.

$$\begin{pmatrix} 8 & -2 & 0 \\ -2 & 10 & -2 \\ 0 & -2 & 12 \end{pmatrix}\begin{pmatrix} i_1 \\ i_2 \\ i_3 \end{pmatrix} = \begin{pmatrix} 4 \\ 12i_x - 5 \\ 5 \end{pmatrix} \qquad (2\text{-}25)$$

From a mathematical point of view, Eq. (2-25) seems to represent a problem with four unknowns but only three equations and would therefore be unsolvable. This is not the case. It will *always* be possible to *express the dependency* in terms of the network variables. In this network it is clear from an examination of Fig. 2–12 that $i_x = i_1$. You see that i_x is given as part of the problem specification, to be the net current through the 4-Ω resistor. But i_1 is, by assignment, the elementary loop current and is also the net current through the 4-Ω resistor, and in the same direction. When this result is substituted into Eq. (2-25), the resulting matrix equation is

$$\begin{pmatrix} 8 & -2 & 0 \\ -2 & 10 & -2 \\ 0 & -2 & 12 \end{pmatrix}\begin{pmatrix} i_1 \\ i_2 \\ i_3 \end{pmatrix} = \begin{pmatrix} 4 \\ 12i_1 - 5 \\ 5 \end{pmatrix} \qquad (2\text{-}26)$$

So now there are clearly only three unknowns and the matrix equation represents a solvable set of simultaneous equations. From a mathematical point of view again, however, the matrix is not in good form, because the forcing function contains one of the unknowns. This is easy to fix, and the same method will be used in every case of controlled sources. To see how the matrix is put in good form, let us write out the second equation from Eq. (2-26):

$$-2i_1 + 10i_2 - 2i_3 = 12i_1 - 5 \qquad (2\text{-}27)$$

This equation is not in good form and to change it simply requires that the $12i_1$ term be moved from the right side to the left, changing its sign. Formally, it is subtracted from both sides of the equation. The result is

$$-14i_1 + 10i_2 - 2i_3 = -5 \qquad (2\text{-}28)$$

This is still the second equation of the set from the matrix, but it has been rearranged to be in standard form. If Eq. (2-28) is reinserted into the matrix format, the result will be

$$\begin{pmatrix} 8 & -2 & 0 \\ -14 & 10 & -2 \\ 0 & -2 & 12 \end{pmatrix} \begin{pmatrix} i_1 \\ i_2 \\ i_3 \end{pmatrix} = \begin{pmatrix} 4 \\ -5 \\ 5 \end{pmatrix} \qquad (2\text{-}29)$$

Equation (2-29) is in the standard form for solution using Cramer's rule or any other method. Looking back, you can see that the adjustment to produce Eq. (2-29) from Eq. (2-26) could have been made directly, without expressing the equation explicitly. The term $12i_1$ can be moved across the equal sign in Eq. (2-26), changing its sign and the 12 is combined with the coefficient of i_1 in the second row.

In general, the approach to solving a network problem with controlled sources, using mesh or nodal analysis, can be summarized as follows:

1. Construct the matrix equation using the normal rules for mesh or nodal analysis, treating the controlled source like any other source and including source transformations if necessary.
2. Study the network and determine how to express the dependency of the controlled sources in terms of the network variables. Insert this dependency into the matrix equation.
3. Move those terms in the forcing function matrix which contain network variables across the equal sign, changing their sign, and combine with appropriate terms in the coefficient matrix.

The following examples illustrate this type of problem.

EXAMPLE 2-17
Find v_{out} of the network of Fig. 2–13.

SOLUTION
Notice that this network is set up with most elements in parallel. Thus it may be easier to use nodal analysis to solve the problem. In Fig. 2–14 the network has been redrawn with the input voltage source transformed to a current source, nodes identified and labeled, the reference selected, and the resistances ex-

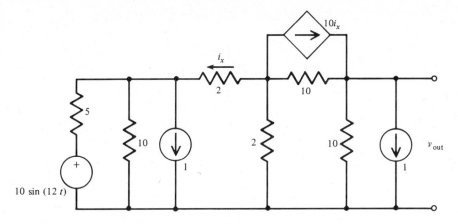

FIGURE 2–13. Network for Example 2-17.

pressed by conductance. The normal rules of nodal analysis give the following matrix:

$$
\begin{pmatrix}
0.8 & -0.5 & 0 \\
-0.5 & 1.1 & -0.1 \\
0 & -0.1 & 0.2
\end{pmatrix}
\begin{pmatrix}
v_1 \\
v_2 \\
v_3
\end{pmatrix}
=
\begin{pmatrix}
2 \sin (12t) - 1 \\
-10i_x \\
10i_x - 1
\end{pmatrix}
$$

The next step is to identify the dependent variable, i_x, in terms of the network variables, v_1, v_2, v_3. This can be done using Ohm's Law. The dependency current is given to be the net current flowing from right to left through the 2-Ω resistor. Since current flows from higher potential to lower potential, this current can be written in terms of the node voltages as

$$
i_x = \frac{v_2 - v_1}{2} = 0.5v_2 - 0.5v_1
$$

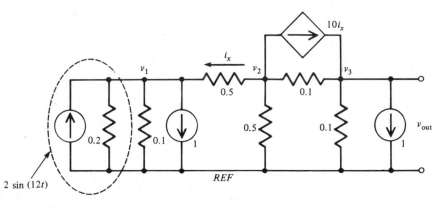

FIGURE 2–14. The network of Fig. 2–13 has been prepared for nodal analysis.

This result is now placed into the matrix equation found above and the result is

$$
\begin{pmatrix}
0.8 & -0.5 & 0 \\
-0.5 & 1.1 & -0.1 \\
0 & -0.1 & 0.2
\end{pmatrix}
\begin{pmatrix}
v_1 \\
v_2 \\
v_3
\end{pmatrix}
=
\begin{pmatrix}
2\sin(12t) - 1 \\
-5v_2 + 5v_1 \\
5v_2 - 5v_1 - 1
\end{pmatrix}
$$

Now the terms in the forcing function that contain v_1 and v_2 are moved across the equal sign, staying in the same row, and combined with the appropriate coefficients. This gives

$$
\begin{pmatrix}
0.8 & -0.5 & 0 \\
-5.5 & 6.1 & -0.1 \\
5 & -5.1 & 0.2
\end{pmatrix}
\begin{pmatrix}
v_1 \\
v_2 \\
v_3
\end{pmatrix}
=
\begin{pmatrix}
2\sin(12t) - 1 \\
0 \\
-1
\end{pmatrix}
$$

Since v_{out} is clearly just v_3, Cramer's rule provides the solution as

$$
v_{\text{out}} = \frac{
\begin{vmatrix}
0.8 & -0.5 & 2\sin(12t) - 1 \\
-5.5 & 6.1 & 0 \\
5 & -5.1 & -1
\end{vmatrix}
}{
\begin{vmatrix}
0.8 & -0.5 & 0 \\
-5.5 & 6.1 & -0.1 \\
5 & -5.1 & 0.2
\end{vmatrix}
}
$$

$$
= 1.19 - 18.28\sin(12t) \qquad \blacksquare
$$

In this example notice that the network *amplified* the input since the input oscillation amplitude was 10 V and the output oscillation amplitude is 18.28 V, a gain of 1.828. The presence of controlled sources is necessary to produce amplification. Notice also that a 180° phase shift has occurred because of the negative sign on the oscillation.

EXAMPLE 2-18

Set up the mesh matrix equation for the network of Fig. 2–15. Put the matrix in final form, ready for application of Cramer's rule or other solving technique.

SOLUTION

Figure 2–16 shows the network with the current source transformed to a voltage source and loop currents assigned. The matrix equation is

$$
\begin{pmatrix}
13 & -7 & 0 & -2 \\
-7 & 15 & -4 & 0 \\
0 & -4 & 11 & -5 \\
-2 & 0 & -5 & 13
\end{pmatrix}
\begin{pmatrix}
i_1 \\
i_2 \\
i_3 \\
i_4
\end{pmatrix}
=
\begin{pmatrix}
6i_x - 3 \\
-4 \\
-10 \\
10 - 6i_x
\end{pmatrix}
$$

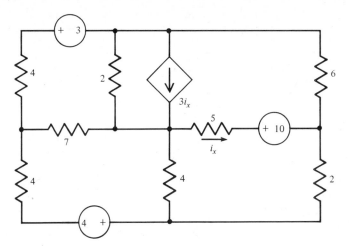

FIGURE 2–15. Network for Example 2-18.

The dependent variable is the *total current* through the 5-Ω resistor. To express this in terms of the network variables, note that i_4 and i_3 both go through this resistor, in opposite directions. Thus the net current is $i_x = i_3 - i_4$. The reason i_3 is positive is that it has the same direction as i_x. The matrix equation is now

$$\begin{pmatrix} 13 & -7 & 0 & -2 \\ -7 & 15 & -4 & 0 \\ 0 & -4 & 11 & -5 \\ -2 & 0 & -5 & 13 \end{pmatrix} \begin{pmatrix} i_1 \\ i_2 \\ i_3 \\ i_4 \end{pmatrix} = \begin{pmatrix} 6i_3 - 6i_4 - 3 \\ -4 \\ -10 \\ 10 - 6i_3 + 6i_4 \end{pmatrix}$$

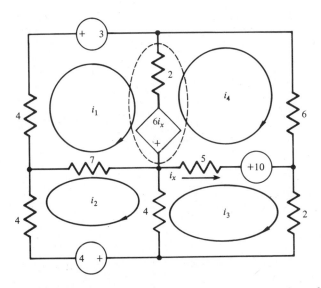

FIGURE 2–16. Network of Fig. 2–15 prepared for mesh analysis.

Now the network variable terms in the forcing function matrix are moved across the equal sign and combined with the appropriate terms in the coefficient matrix. This gives the final result,

$$\begin{pmatrix} 13 & -7 & -6 & 4 \\ -7 & 15 & -4 & 0 \\ 0 & -4 & 11 & -5 \\ -2 & 0 & 1 & 7 \end{pmatrix} \begin{pmatrix} i_1 \\ i_2 \\ i_3 \\ i_4 \end{pmatrix} = \begin{pmatrix} -3 \\ -4 \\ -10 \\ 10 \end{pmatrix} \qquad \blacksquare$$

2-5

APPLICATIONS

Mesh and nodal analysis of resistive networks are very useful tools for solving many types of problems in electricity and electronics. In later chapters you will see that the same principles can be applied to networks with capacitors and inductors. In this section an application of mesh and nodal analysis will be presented.

Selection of Method.   There is often a question of which method of analysis to pick for solving a particular network problem, mesh or nodal. The fact is that either method can usually be used to solve a network. The selection of one over the other often then reduces to using that method which provides the simplest path to the solution. For example, the method that gives the fewest unknowns is often a factor in selection. Another factor can be the observation of which method provides the solution most directly. If a voltage across some element is desired, nodal analysis is often preferred since it solves for voltages directly. If currents are the desired solutions, mesh analysis is probably the best choice.

In general, if a network appears in the form of mostly serial loops of elements, mesh analysis will be easier, and if laid out in mostly parallel arrangements of elements, nodal analysis will be easier.

Current Sources with Mesh Analysis.   When the rules for mesh analysis were given it was stated that any current sources should be transformed to voltage sources. This is true if the rules are strictly followed. However, it is possible to use mesh analysis in networks with currents sources. When this happens the system will simply have *one less unknown* than there are mesh current loops. Therefore, the matrix equation must be reduced in order by one by inserting the known loop current, from the given source, into the equations and simplifying. The contribution from the given current will simply add or subtract to terms in the forcing function matrix.

Voltage Sources with Nodal Analysis. If the rules for nodal matrix construction are strictly followed, no voltage sources can be in the network. Any voltage sources must be converted to current sources. However, it is possible to have voltage sources in the network and still perform nodal analysis. All the voltage source means is that the system has one less unknown than would be predicted by nodal analysis. This is because the given voltage source fixes the voltage between two nodes between which it is connected. In this case the matrix equation must be adjusted by reducing the order of the matrix by substituting the known node voltage. The known source will simply add or subtract to terms in the forcing function matrix.

2-5.1 Thévenin and Norton Equivalent Circuits

One of the difficulties with using the Thévenin and Norton equivalent circuit concept in analysis is that it is often difficult to calculate the equivalent circuit values in complex circuits. Mesh and nodal analysis simplify this calculation.

Thévenin Equivalent Circuit. *The Thévenin equivalent circuit law for resistive networks states: Any two terminals connected to a resistive network may be replaced by a single voltage source, v_{TH}, in series with a single resistance, R_{TH}.*

Figure 2–17 illustrates schematically the meaning of this law. All electrical characteristics measured between terminals a and b of the two systems will be

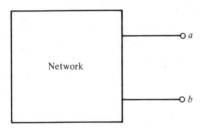

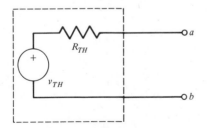

FIGURE 2–17. Thévenin's law allows the replacement of an entire resistive network by a single voltage source and resistor.

identical. This is true no matter what types of sources are contained in the original network. The only restriction is that the original network contain only resistors, that is, no capacitors, inductors, or transformers. The values of the Thévenin equivalent circuit elements are found as follows:

v_{TH}.   *The Thévenin voltage is the open-circuit voltage that exists between terminals a and b of the original network.*

R_{TH}.   *The Thévenin resistance is the resistance between terminals a and b of the original network, with all* real *sources replaced by their internal resistance.*

Thus one can find the Thévenin voltage by a direct application of mesh or nodal analysis to find the voltage from *a* to *b*. In the case of the Thévenin resistance, ideal voltage sources are replaced by a short and ideal current sources are replaced by an open. Controlled sources are *not real sources* but models of active devices and must be left in the network, even to find the resistance. In such cases it is impossible to find the equivalent resistance by series and parallel reductions, as the following examples illustrate.

EXAMPLE 2-19

Given the network of Fig. 2–18, find the Thévenin equivalent circuit for the terminals labeled *a* and *b*.

SOLUTION

Let us use mesh analysis to solve this problem. First, the 5-A current source shunted by 4 Ω is transformed into a 20-V source in series with the 4-Ω resistor. Figure 2–19 shows the network prepared for mesh analysis to find the Thévenin voltage. Note that the 8-Ω resistor is not included in this analysis because no current passes through this element. Then the Thévenin voltage is found as $v_{TH} = v_{ab} = 4i_2$. The matrix equation is

$$\begin{pmatrix} 8 & -4 \\ -4 & 10 \end{pmatrix} \begin{pmatrix} i_1 \\ i_2 \end{pmatrix} = \begin{pmatrix} -14 \\ 20 - 2 \sin(3t) \end{pmatrix}$$

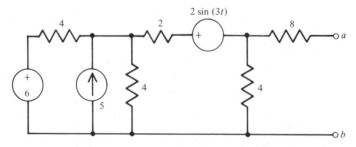

FIGURE 2–18.   Network for Example 2-19.

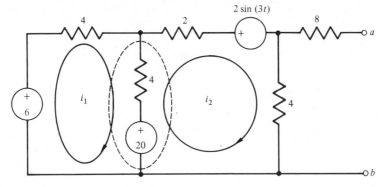

FIGURE 2–19. The network of Fig. 2–18 has been prepared for mesh analysis to find the Thévenin voltage.

Therefore, i_2 is found as

$$i_2 = \frac{\begin{vmatrix} 8 & -14 \\ -4 & 20 - 2\sin(3t) \end{vmatrix}}{\begin{vmatrix} 8 & -4 \\ -4 & 10 \end{vmatrix}} = 1.625 - 0.25\sin(3t)$$

So now the Thévenin voltage is

$$v_{TH} = 6.5 - \sin(3t)$$

To find the Thévenin resistance the sources are replaced by their internal resistance and the resulting network is shown in Fig. 2–20. This network can be simplified by series and parallel combinations of resistors until one resistance remains. This is the Thévenin resistance. The result is

$$R_{TH} = 10\ \Omega$$

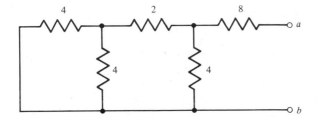

FIGURE 2–20. Network of Fig. 2–18 for finding the Thévenin resistance.

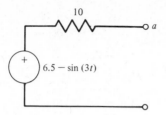

FIGURE 2–21. This is the Thévenin equivalent circuit of the network of Fig. 2–18.

Figure 2–21 shows the Thévenin equivalent circuit of Fig. 2–18. The polarity of the source was determined by the fact that current i_2 flows into the upper terminal of the 4-Ω resistor. ∎

The next example illustrates the effect of a controlled source in the network, as well as showing how the equivalent circuit can be used to simplify a problem.

EXAMPLE 2-20

Given the network of Fig. 2–22, find the load current for the different loads shown.

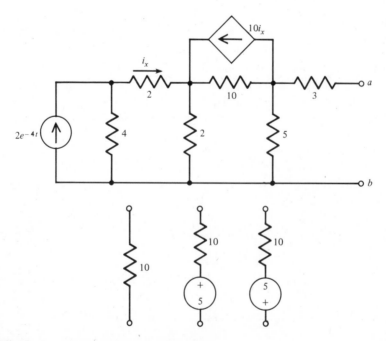

FIGURE 2–22. Network for Example 2-20.

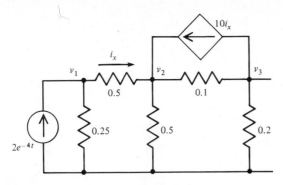

FIGURE 2–23. Network of Fig. 2–22 set up for finding the Thévenin voltage by nodal analysis.

SOLUTION

In general, since there are three different loads to be evaluated, it would be necessary to connect each load in turn and solve the network problem to find the current. However, by using the Thévenin equivalent circuit this problem will be much simpler since it will be a matter of computing the current of a single loop. v_{TH} will be the voltage across the 5-Ω resistor. Figure 2–23 shows the network set up to find v_{TH} using nodal analysis. Note that $v_{TH} = v_3$. Of course, the 3-Ω resistor has been left off since no current goes through this component. The matrix equation is

$$\begin{pmatrix} 0.75 & -0.5 & 0 \\ -0.5 & 1.1 & -0.1 \\ 0 & -0.1 & 0.3 \end{pmatrix} \begin{pmatrix} v_1 \\ v_2 \\ v_3 \end{pmatrix} = \begin{pmatrix} 2e^{-4t} \\ 10i_x \\ -10i_x \end{pmatrix}$$

An examination of Fig. 2–23 shows that the dependent current is given by

$$i_x = \frac{v_1 - v_2}{2} = 0.5v_1 - 0.5v_2$$

When this relationship is inserted into the matrix equation and the necessary rearrangements are made, the final matrix equation is

$$\begin{pmatrix} 0.75 & -0.5 & 0 \\ -5.5 & 6.1 & -0.1 \\ +5 & -5.1 & 0.3 \end{pmatrix} \begin{pmatrix} v_1 \\ v_2 \\ v_3 \end{pmatrix} = \begin{pmatrix} 2e^{-4t} \\ 0 \\ 0 \end{pmatrix}$$

Solving for v_3 will find v_{TH}:

$$v_{TH} = \frac{\begin{vmatrix} 0.75 & -0.5 & 2e^{-4t} \\ -5.5 & 6.1 & 0 \\ +5 & -5.1 & 0 \end{vmatrix}}{\begin{vmatrix} 0.75 & -0.5 & 0 \\ -5.5 & 6.1 & -0.1 \\ +5 & -5.1 & 0.3 \end{vmatrix}} = -11.8e^{-4t}$$

To find R_{TH}, the real source in Fig. 2–22 is set equal to its internal resistance and the resulting network is shown in Fig. 2–24. Note that we cannot find the resistance from a to b by series and parallel reductions because of the presence of the controlled source. *In these cases* the equivalent resistance is found by connecting a source to a and b and calculating the response of the circuit. If a voltage source is connected, then the current is calculated which, with the source, will give the equivalent resistance by Ohm's law. If a current source is connected, the calculated voltage across the source will determine the resistance. In this problem we apply a current source and calculate the voltage using nodal analysis. Figure 2–25 shows the resulting network set up for nodal analysis. The matrix equation is

$$\begin{pmatrix} 0.767 & -0.1 & 0 \\ -0.1 & 0.633 & -0.333 \\ 0 & -0.333 & 0.333 \end{pmatrix} \begin{pmatrix} v_1 \\ v_2 \\ v_3 \end{pmatrix} = \begin{pmatrix} 10i_x \\ -10i_x \\ i \end{pmatrix}$$

The dependent variable is $i_x = -0.167v_1$. It is negative because the current is defined by the problem to have the direction given, which implies current flows from the reference to v_1, and that the reference is at a higher potential. After insertion and rearrangement, the matrix equation is

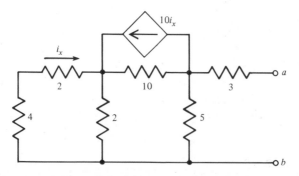

FIGURE 2–24. To find the Thévenin resistance of the network of Fig. 2–22 the controlled source must be left in the network.

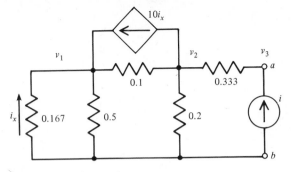

FIGURE 2–25. A current source is applied to the network of Fig. 2–24 to find the Thévenin resistance.

$$\begin{pmatrix} 2.433 & -0.1 & 0 \\ -1.767 & 0.633 & -0.333 \\ 0 & -0.333 & 0.333 \end{pmatrix} \begin{pmatrix} v_1 \\ v_2 \\ v_3 \end{pmatrix} = \begin{pmatrix} 0 \\ 0 \\ i \end{pmatrix}$$

This can be solved by Cramer's rule to find v_3 to be

$$v_3 = 7.4i$$

Then

$$R_{\text{TH}} = \frac{v_3}{i} = 7.4 \ \Omega$$

The Thévenin equivalent circuit is shown in Fig. 2–26. Now each load can be connected and the current quickly calculated by Ohm's law, as illustrated in Fig. 2–27 for one of the loads. The results are

$$\text{Case 1:} \quad i_L = -1.47e^{-4t}$$

$$\text{Case 2:} \quad i_L = -1.47e^{-4t} - 0.287$$

$$\text{Case 3:} \quad i_L = -1.47e^{-4t} + 0.287 \qquad \blacksquare$$

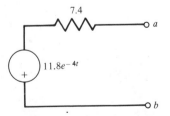

FIGURE 2–26. Thévenin equivalent circuit of the network of Fig. 2–22.

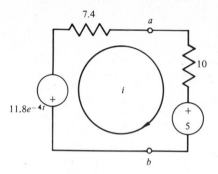

FIGURE 2–27. Thévenin equivalent circuit being used to find the effect of a load on the network of Fig. 2–22.

Norton Equivalent Circuit. *The Norton equivalent circuit for a linear resistive network is defined as follows: Any two terminals connected to a network may be replaced by a single current source, i_N, shunted by a single resistor, R_N.*

The sources in the network may depend on time or be controlled sources. Figure 2–28 illustrates schematically the nature of this equivalent circuit law. All electrical characteristics of the original network measured between terminals a and b are copied by the Norton equivalent circuit. The values of the equivalent circuit are found according to the following rules;

I_N: *The Norton equivalent current is that current which flows through a short circuit placed between terminals a and b.*

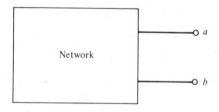

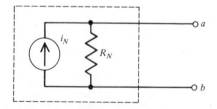

FIGURE 2–28. The Norton equivalent circuit can replace an entire network.

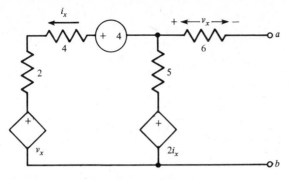

FIGURE 2–29. Network for Example 2-21.

R_N: *The Norton resistance is the resistance between terminals a and b if all real sources have been replaced by their internal resistance (same as the Thévenin equivalent resistance).*

Actually, you should recognize that the Thévenin and Norton equivalent circuits are merely the transformations of each other. Thus, if you have found one, the other can be found by the transformation of a current source to a voltage source, and vice versa. In equation form we have

$$R_N = R_{TH} \qquad \text{and} \qquad i_N = \frac{v_{TH}}{R_{TH}}$$

EXAMPLE 2-21

Find the Norton equivalent circuit for the network given in Fig. 2–29.

SOLUTION

We will use mesh analysis to find i_N for this problem. When a short is placed from *a* to *b* and the loops are identified, the result is a network as shown in Fig. 2–30. The matrix equation is quickly found as

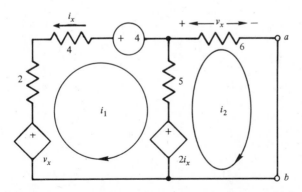

FIGURE 2–30. Network of Fig. 2–29 as set up to find the Norton current using mesh analysis.

$$\begin{pmatrix} 11 & -5 \\ -5 & 11 \end{pmatrix}\begin{pmatrix} i_1 \\ i_2 \end{pmatrix} = \begin{pmatrix} v_x - 2i_x - 4 \\ 2i_x \end{pmatrix}$$

From an examination of the network it is clear that the dependency is $i_x = -i_1$ and $v_x = 6i_2$. Then the matrix equation becomes

$$\begin{pmatrix} 9 & -11 \\ -3 & 11 \end{pmatrix}\begin{pmatrix} i_1 \\ i_2 \end{pmatrix} = \begin{pmatrix} -4 \\ 0 \end{pmatrix}$$

The Norton equivalent current will simply be i_2, so solving for this current,

$$i_N = i_2 = \frac{\begin{vmatrix} 9 & -4 \\ -3 & 0 \end{vmatrix}}{\begin{vmatrix} 9 & -11 \\ -3 & 11 \end{vmatrix}} = -0.18 \text{ A}$$

To find the resistance the 4-V real source is replaced by its internal resistance of a short circuit and the network of Fig. 2–31 results. As before, we must place a source on terminals a and b and calculate the response to determine the resistance. In this case a voltage source will be used, as shown in Fig. 2–32. The matrix equation is

$$\begin{pmatrix} 11 & -5 \\ -5 & 11 \end{pmatrix}\begin{pmatrix} i_1 \\ i_2 \end{pmatrix} = \begin{pmatrix} v_x - 2i_x \\ 2i_x - v \end{pmatrix}$$

Then $i_x = -i_1$ and $v_x = 6i_2$, so the matrix equation becomes

$$\begin{pmatrix} 9 & -11 \\ -3 & 11 \end{pmatrix}\begin{pmatrix} i_1 \\ i_2 \end{pmatrix} = \begin{pmatrix} 0 \\ -v \end{pmatrix}$$

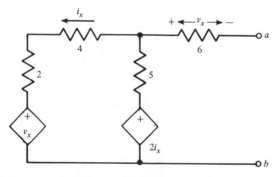

FIGURE 2–31. Network of Fig. 2–29 set up for finding the Norton resistance.

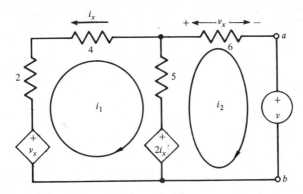

FIGURE 2–32. The Norton resistance will be found by applying a voltage source and calculating the current.

The current of the applied source is i_2, so we solve for this quantity using Cramer's rule:

$$i_2 = -0.136v$$

The negative sign simply notes that the current really comes out of the positive terminal of the source. Thus the Norton resistance is

$$i_N = \frac{v}{i_2} = 7.4 \ \Omega$$

The Norton equivalent circuit is shown in Fig. 2–33. ■

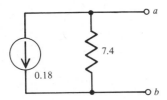

FIGURE 2–33. This is the Norton equivalent circuit of the network in Fig. 2–29.

2-5.2 Computer Applications

The solution of a resistive network problem involves two steps:

1. Develop the matrix equation.
2. Solve the equation for the unknown or unknowns desired.

The computer can be employed to help accomplish both of these steps but is particularly useful for the second.

Determinant Evaluation.   Most user-oriented and time-sharing computer systems have a math package of prepared programs available in a library of programs. Such a package will contain a program for the evaluation of determinants. In such a case, following the development of the matrix equation, this program can be used to facilitate solution of the unknowns using Cramer's rule. Typically, however, the determinant must contain *only* numbers, that is, no time-dependent functions. If time-dependent functions are present, Laplace expansion can be used to isolate these functions from the determinants to be evaluated. The following example illustrates this process.

EXAMPLE 2-22

The following matrix equation resulted from an application of mesh analysis to a network problem.

$$\begin{pmatrix} 4 & -2 & 0 & -1 \\ -2 & 10 & -4 & 0 \\ 0 & -4 & 12 & -2 \\ -1 & 0 & -2 & 18 \end{pmatrix} \begin{pmatrix} i_1 \\ i_2 \\ i_3 \\ i_4 \end{pmatrix} = \begin{pmatrix} -5 \\ 4 + 2 \cos (2t) \\ 0 \\ -1 \end{pmatrix}$$

Set up the solution for i_2 using Cramer's rule such that a computer determinant evaluation program can be used.

SOLUTION

Cramer's rule gives the solution for i_2 as

$$i_2 = \frac{\begin{vmatrix} 4 & -5 & 0 & -1 \\ -2 & 4 + 2 \cos (2t) & -4 & 0 \\ 0 & 0 & 12 & -2 \\ -1 & -1 & -2 & 18 \end{vmatrix}}{\begin{vmatrix} 4 & -2 & 0 & -1 \\ -2 & 10 & -4 & 0 \\ 0 & -4 & 12 & -2 \\ -1 & 0 & -2 & 18 \end{vmatrix}}$$

The determinant in the denominator can be evaluated directly using the computer program since it contains only numbers. The numerator cannot because of the term in $\cos (2t)$. Now, Laplace expansion will be used to isolate this term. First, zero generation is used to produce another zero in column 2 by subtracting five times row 4 from row 1, leaving

$$\begin{vmatrix} 9 & 0 & 10 & -91 \\ -2 & 4 + 2\cos(2t) & -4 & 0 \\ 0 & 0 & 12 & -2 \\ -1 & -1 & -2 & 18 \end{vmatrix}$$

Now Laplace expansion on column 2 yields

$$[4 + 2\cos(2t)]\begin{vmatrix} 9 & 10 & -91 \\ 0 & 12 & -2 \\ -1 & -2 & 18 \end{vmatrix} - \begin{vmatrix} 9 & 10 & -91 \\ -2 & -4 & 0 \\ 0 & 12 & -2 \end{vmatrix}$$

Each of these determinants can now be evaluated by the computer since they contain only numbers. ∎

Simultaneous Equations. Many math packages contain programs which directly solve a set of simultaneous equations. The input in this case are the coefficient matrix elements and the forcing function matrix elements. The computer then calculates the unknowns. If time functions are present, it will be necessary to use Cramer's Rule for evaluation since the programs will not accept time functions as the forcing functions.

Plotting. As mentioned in Chapter 1, the finding of a solution is often only a means to the end. The next step is to interpret the result and draw conclusions about network behavior. Often changes in the network are deduced as a result of the solutions. A plot or graph of the solution as a function of time is often used to aid in an understanding of what the solution means. Many math packages contain plotting routines that enable a given function of time to be plotted on a printer over specified intervals of time.

SUMMARY

The primary objective of this chapter was to introduce the powerful matrix construction methods of setting up the solutions to resistive network problems, mesh and nodal analysis. The following points were covered in the task of covering this objective:

1. The method of setting up simultaneous equations for a network using KVL in a general sense was presented. This method, which is also called *loop analysis*, uses KVL to set up the equations and KCL with Ohm's law to simplify the equations.
2. The use of KCL, also called *node analysis*, was presented as a technique that uses KCL to generate the equations and KCL with Ohm's law to simplify the equations.

3. An extension of loop analysis is shown to lead to mesh analysis, which allows the matrix representation of the simultaneous equations for network currents to be written down, almost by inspection.

4. The method of node analysis is shown to lead to nodal analysis, which allows the matrix representation of the simultaneous equations for network voltages to be written down from the network, almost by inspection.

5. A review was presented of matrix algebra, illustrating how a set of simultaneous equations can be written as a matrix equation. The method of multiplying two matrices was presented.

6. The method of solving a set of simultaneous equations by hand, called Cramer's rule, was defined and applied to network problems.

7. A review was presented of Laplace expansion and zero generation as powerful aids in the evaluation of higher-order (>3) determinants.

8. The mesh and nodal approaches were applied to the problem of finding the Thévenin and Norton equivalent circuits of networks.

PROBLEMS

2–1   Using KVL loop analysis, find the current magnitude and direction through each resistor in the networks of Fig. 2–34.

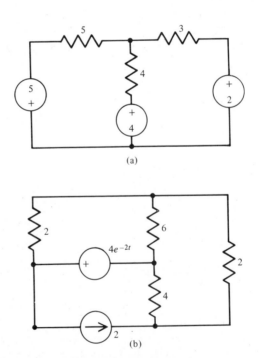

FIGURE 2–34. Network for Problem 2–1.

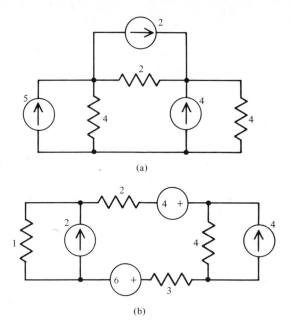

(a)

(b)

FIGURE 2–35. Network for Problem 2–2.

2–2   Using KCL node analysis, find the voltage polarity and magnitude across each resistor in the networks of Fig. 2–35.

2–3   Given the matrix

$$\mathbf{A} = \begin{pmatrix} 0 & -4 & 2 & 1 \\ 2 & 1 & -4 & 3 \\ 0 & -8 & 6 & 1 \end{pmatrix}$$

specify the values of the elements a_{12}, a_{21}, a_{34}, and a_{23}.

2–4   An inventory of capacitors is categorized by value and working voltage rating. Construct an inventory matrix for the following data, where the rows are capacity and the columns are working voltage. Row 1 is to be the lowest capacity and column 1 the lowest voltage.

0.01 µF:	5 at 500 V;	10 at 1 kV;	3 at 2 kV
0.5 µF:	0 at 500 V;	13 at 1 kV;	6 at 2 kV
200 pF:	4 at 500 V;	0 at 1 kV;	7 at 2 kV
1 µF:	10 at 500 V;	0 at 1 kV;	1 at 2 kV

2–5   Find the products of the following matrices.

(a)
$$\begin{pmatrix} 2 & -1 \\ 4 & 2 \end{pmatrix} \begin{pmatrix} 2 & 0 \\ 2 & 1 \end{pmatrix}$$

(b)
$$\begin{pmatrix} 3 & -1 & 4 \\ 2 & 0 & 1 \\ 1 & 1 & 1 \end{pmatrix} \begin{pmatrix} 2 & 1 \\ 3 & 1 \\ -4 & 2 \end{pmatrix}$$

(c)
$$\begin{pmatrix} 3 & -2 \\ 4 & 3 \\ 6 & 9 \end{pmatrix} \begin{pmatrix} 2 & 1 \\ 1 & 4 \end{pmatrix}$$

2–6   Find the values of x and y by multiplying out the following matrix equation.

$$\begin{pmatrix} 3 & x \\ 2 & y \end{pmatrix} \begin{pmatrix} 1 \\ 2 \end{pmatrix} + \begin{pmatrix} 4 \\ 7 \end{pmatrix} = \begin{pmatrix} -2 \\ 1 \end{pmatrix}$$

2–7   Express the following set of simultaneous equations as a matrix equation.

$$\begin{aligned}
6x - 2y - 4z + 2w &= -4 \\
2y + 4z - w &= 0 \\
3x \qquad\quad - 2z + 4w &= -7 \\
-6x - 4y + z - 3w &= 14
\end{aligned}$$

2–8   Find the solutions of the following matrix equation using Cramer's rule.

$$\begin{pmatrix} 8 & -4 \\ 6 & 10 \end{pmatrix} \begin{pmatrix} x \\ y \end{pmatrix} = \begin{pmatrix} 2 \sin (3t) \\ -4 \end{pmatrix}$$

2–9   Evaluate the following determinants.

(a) $\begin{vmatrix} 4 & -2 \\ 6 & -3 \end{vmatrix}$ **(b)** $\begin{vmatrix} 2 & 0 & 7 \\ -1 & 4 & -1 \\ 2 & 2 & 1 \end{vmatrix}$

(c) $\begin{vmatrix} 2 \cos (2t) & 3 & -1 \\ 7 & 4 & 4e^{-3t} \\ 0 & -1 & 4 \end{vmatrix}$

2–10   For the following determinant, find the cofactors R_{31}, R_{23}, and R_{42}.

$$\begin{vmatrix} 2 & -1 & 0 & 4 \\ 2 & 1 & 1 & 4 \\ -4 & 5 & 0 & 2 \\ 1 & 0 & 4 & -1 \end{vmatrix}$$

2–11   Evaluate the determinant of Problem 2–10 by Laplace expansion along column 3.

2–12   Evaluate the determinant of Problem 2–10 along row 4, but use zero generation to reduce to one 3 × 3 determinant.

2–13   Solve for x, y, z, and w in Problem 2–7 using Cramer's rule, Laplace expansion, and zero generation.

2–14   Evaluate the following determinant.

$$\begin{vmatrix} -2 & 0 & -4 & 1 & 0 \\ 2 & 0 & 2 & 3 & -1 \\ 1 & 4 & -6 & 4 & -2 \\ -3 & 2 & -4 & 0 & -2 \\ 1 & 2 & -1 & 4 & 0 \end{vmatrix}$$

2–15   Use mesh analysis to solve the networks of Fig. 2–36. A solution is the values of all elementary loop currents.

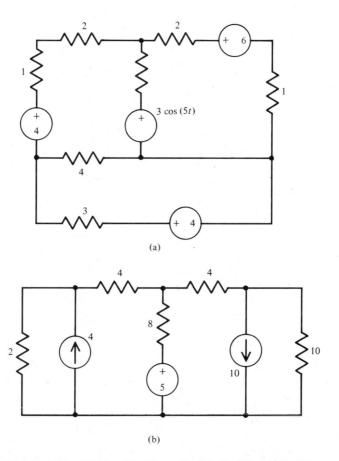

(a)

(b)

FIGURE 2–36.   Network for Problems 2–15, 2–16, and 2–22.

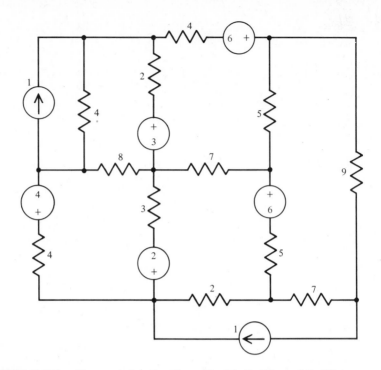

FIGURE 2–37.   Network for Problems 2–17, 2–28, and 2–29.

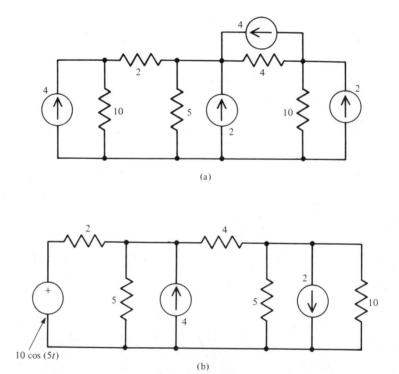

(a)

(b)

FIGURE 2–38.   Network for Problems 2–18, 2–19, and 2–23.

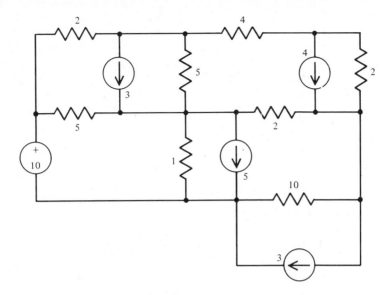

FIGURE 2–39. Network for Problems 2–20 and 2–30.

2–16 Find the voltage across the 2-Ω resistor in Fig. 2–36b.

2–17 For the network of Fig. 2–37, find the mesh matrix equation for the network variables.

2–18 Use nodal analysis to solve the networks of Fig. 2–38. A solution is the values of all node voltages.

2–19 Find the current through the 2-Ω resistor in the network of Fig. 2–38b.

2–20 Use nodal analysis to find the matrix equation for the network given in Fig. 2–39.

2–21 Find the output voltage of the network given in Fig. 2–40. What is the gain in the oscillating part of the output voltage?

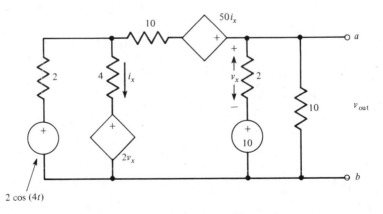

FIGURE 2–40. Network for Problems 2–21 and 2–24.

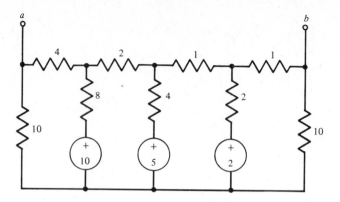

FIGURE 2–41. Network for Problem 2–25.

2–22 Find the Thévenin equivalent circuit for terminals across the 10-Ω resistor in Fig. 2–36b. What is the Norton equivalent circuit?

2–23 Find the Norton equivalent circuit directly for terminals across the 10-Ω resistor in Fig. 2–38b. What is the Thévenin equivalent circuit?

2–24 Find the Thévenin equivalent circuit at terminals a and b of the network in Fig. 2–40.

2–25 Find the Thévenin equivalent circuit for the network of Fig. 2–41 for terminals labeled a and b.

2–26 For the network of Fig. 2–42 plot the dc power and signal power versus load resistance connected to terminals a and b. At what resistance does the signal power reach a peak? The signal is the oscillating part of the voltage and current. (*Hint:* Replace the network by an equivalent circuit at terminals a and b.)

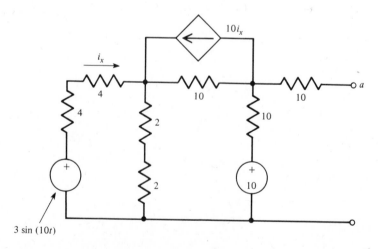

FIGURE 2–42. Network for Problem 2–26.

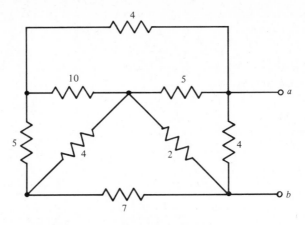

FIGURE 2–43.   Network for Problem 2–27.

2–27   Find the resistance from a to b for the network of Fig. 2–43.

COMPUTER PROBLEMS

2–28   Use Cramer's rule and a determinant evaluation program to find the current through the 9-Ω resistor in the network of Fig. 2–37.

2–29   Use a simultaneous equation solving program to find the solution of all currents in the network of Fig. 2–37, using mesh analysis.

2–30   Use a simultaneous equation solving program to find the node voltages for the network of Fig. 2–39.

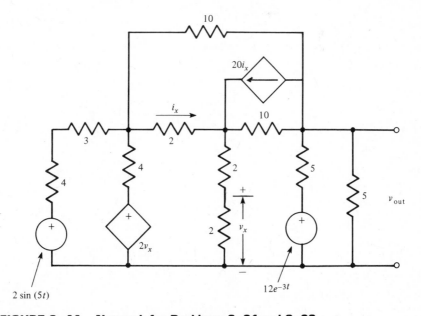

FIGURE 2–44.   Network for Problems 2–31 and 2–32.

2–31   Find the output voltage of the network in Fig. 2–44. Plot the output voltage versus time.

2–32   Find the output resistance of the network of Fig. 2–44. The output resistance is just the Thévenin resistance across the output terminals.

CHAPTER 3

Networks with Capacitors and Inductors

Objectives

The primary objective of this chapter is to eliminate the restriction of allowing only resistors in the networks to be solved. After study of this chapter and completion of the problems at the end of the chapter, you will be able to:

1. Define the concept of time origin of network problems and initial conditions on capacitors and inductors.
2. Develop the integrodifferential equations for any network with resistors, capactors, inductors, and transformers using KVL.
3. Develop the integrodifferential equations for any network with resistors, capacitors, inductors, and transformers using KCL.
4. Show how network integrodifferential equations can be turned in to first- or second-order differential equations.
5. Find the solution of a linear first-order differential equation with constant coefficients.
6. Find the solution of a linear second-order differential equation with constant coefficients.

NETWORK EQUATIONS WITH CAPACITORS AND INDUCTORS

In Chapter 2 you learned how to solve a linear network that contained only resistors as the component elements. In this chapter that restriction will be eliminated and any linear element will be usable in the network. Thus capacitors, inductors, and transformers may now appear in the networks. In this section of the chapter you will learn how to use KVL and KCL to develop equations for networks that contain capacitors and inductors. The previous methods of mesh and nodal analysis will not work without the modifications to be learned in the next chapter. The resulting equations will be found to contain both integrals and derivatives of the network variables. Such equations are called *integrodifferential* equations. The last section of this chapter will present a method of solving these equations by converting them to *differential* equations and finding the solutions of these equations directly.

3-1.1 KVL and KCL

Kirchhoff's voltage and current laws are fundamental relationships of electrical networks. They can be used to set up equations for any network, regardless of components, sources, time dependence, or even linearity. For this reason it is clear that these laws can be used to set up the equations for networks that contain capacitors and inductors. The only problem, then, is how to express the relationship between current and voltage for these components.

In Chapter 1 relations between current and voltage for these components were presented as

$$i(t) = C \frac{dv}{dt} \tag{1-6}$$

for the relationship between current and voltage for a capacitor and

$$v(t) = L \frac{di}{dt} \tag{1-8}$$

for the inductor. This gives the starting point for use of KVL and KCL to develop equations for networks with these components.

KVL. The network of Fig. 3–1 is a series combination of elements, including a capacitor and inductor. For such a series arrangement the most suitable approach is KVL, since this law involves the summation of voltages around a closed loop. In anticipation of this, a loop current, clockwise for convenience, has been drawn around the network. Each element will have a voltage associated

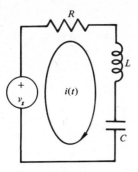

FIGURE 3–1. Network with a capacitor and inductor for illustrating the application of KVL.

with it and KVL specifies that the sum of these voltages will give zero. Thus an equation can be written:

$$v_s + v_R + v_C + v_L = 0 \qquad (3\text{-}1)$$

where

v_s = given source voltage

v_R, v_C, v_L = voltage drops across the resistor, capacitor, and inductor

Equation (3-1) is one equation in three unknowns since the voltage drops across the components are unknown. To reduce this to one equation in one unknown, each voltage drop will be expressed in terms of the loop current.

For the resistor this is done using Ohm's law,

$$v_R = Ri(t) \qquad (3\text{-}2)$$

where the polarity will be positive on the side of the resistor which the current enters.

For the inductor the voltage drop can be found from Eq. (1-8) to be

$$v_L = L\frac{di}{dt} \qquad (3\text{-}3)$$

Again, the positive of this voltage will be on the side of the inductor from which the loop current enters the inductor.

What remains, then, is to express the voltage drop across the capacitor in terms of the current. Equation (1-6) gives the current as a function of the voltage, but this will not help in the present problem. However, by *integrating* Eq. (1-6) it will be possible to derive an equation for the capacitor voltage in terms of the current. Let us do this step by step so that it will be clear where each term comes from. First solve Eq. (1-6) for dv_C in terms of $i(t)$,

$$dv_C = \frac{1}{C} i(t) \, dt$$

and then integrate both sides,

$$\int_{t_0}^{t} dv_C = \frac{1}{C} \int_{t_0}^{t} i(\tau) \, d\tau \qquad (3\text{-}4)$$

where

$$t, t_0 = \text{limits of integration}$$

$$\tau = \text{variable of integration}$$

To be completely specific in the problem it is necessary to provide limits of integration. The lower limit, t_0, specifies the moment from which consideration of the problem begins, that is, the initial starting time of the problem to be solved. This is called the *time origin* of the problem. The upper limit denotes the moment in time, after t_0, when the value of the voltage drop of the capacitor is to be determined. Equation (3-4) can now be integrated to find an expression for the voltage drop across the capacitor at time t,

$$v_C(t) - v_C(t_0) = \frac{1}{C} \int_{t_0}^{t} i(\tau) \, d\tau \qquad (3\text{-}5)$$

Now, Eq. (3-5) can be solved for the voltage drop across the capacitor itself as

$$v_C(t) = v_C(t_0) + \frac{1}{C} \int_{t_0}^{t} i(\tau) \, d\tau \qquad (3\text{-}6)$$

The physical significance of this equation is very important. The term $v_C(t_0)$ is the voltage on the capacitor at time t_0, which is the time at which the problem "started," that is, the initial time from which our consideration of the voltage drop of the capacitor began. This voltage is called an *initial condition* on the capacitor. Integral relationships always give such initial conditions since an integral specifies the present value of a variable based on the history of another variable. The integral part of Eq. (3-6) states that the accumulation of voltage on the capacitor is the result of current flowing through the capacitor from the starting time, t_0, to the present time, t. This means that the capacitor voltage *cannot* change instantaneously. We say that the voltage on a capacitor is *continuous*.

The previous considerations lead to the statement of two facts about networks with capacitors:

1. For problems with capacitors a *time origin* must be specified—as that time after which equations developed will specify the values of network variables. It is often convenient to choose the time origin as zero, $t_0 = 0$, but you should remember that this is not necessary, just convenient.
2. Whenever capacitors are present in a network the value of the voltage on the capacitor at the time origin must be determined and included in the equations. This is called the initial condition of the capacitor.

The first of these requirements will be met in future problems by using a "switch" to turn on the network at a specified time, usually $t = 0$, and thereby establish a time origin for the problem. The second will be met by always determining the initial voltage on capacitors in the network before attempting to solve the problem.

Returning now to the problem of Fig. 3–1, the voltage drop of the capacitor can be written from the general result of Eq. (3-6). The time origin will be taken as $t_0 = 0$; then

$$v_C(t) = v_C(0) + \frac{1}{C} \int_0^t i(\tau)\, d\tau \tag{3-7}$$

where the positive of the voltage-drop integral term will be on the side of the capacitor from which the loop current enters. The polarity of the initial condition voltage will be whatever it turns out to be from analysis of the initial condition.

Equation (3-1) can now be written as one equation in one unknown by using the relations given in Eqs. (3-2), (3-3), and (3-7). The result is

$$v_s - Ri(t) - L\frac{di}{dt} - v_C(0) - \frac{1}{C} \int_0^t i(\tau)\, d\tau = 0 \tag{3-8}$$

Equation (3-8) is called a linear integrodifferential equation for the current $i(t)$. This equation demonstrates that the introduction of capacitors and inductors into networks results in integrals and/or derivatives in the equations to be solved. These equations can be found using KVL and the $i(t)$ versus $v(t)$ relationships of the components as shown above. The following example illustrates the process of finding the equation of a network.

EXAMPLE 3-1
In the network of Fig. 3–2, the switch moves from position a to position b at time $t = 0$. Find the equation for the loop current.

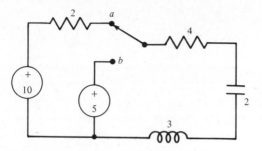

FIGURE 3–2. Network for Example 3-1.

SOLUTION

The time origin of the problem is defined as $t_0 = 0$. After this time (i.e., for $t > 0$) the network will have the form of Fig. 3–3, and this is the network to be solved. The loop equation from KVL will be

$$5 + v_{4\Omega} + v_{2F} + v_{3H} = 0$$

From Ohm's Law, $v_{4\Omega} = 4i(t)$ and for the inductor Eq. (1-8) gives $v_{3H} = 3(di/dt)$. For the capacitor Eq. (3-7) can be used to write

$$v_{2F} = v_{2F}(0) + \tfrac{1}{2} \int_0^t i(\tau)\, d\tau$$

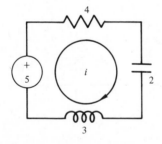

FIGURE 3–3. Network to be solved in Example 3-1. A clockwise loop current has been assigned.

To find the initial condition on the capacitor, $v_{2F}(0)$, the network must be examined *just before* the switch was moved from a to b (i.e., in position a). Figure 3–4 shows the network in position a, and the assumption is that the network has been in this state for a long time. Clearly, there can be no current since the capacitor blocks dc current and there is only a dc source. If there is no current there can be no voltage across the resistors. Since one side of the resistors is at $+10$ V, the other side must be at $+10$ V. This means that there must be a

CAPACITORS AND INDUCTORS

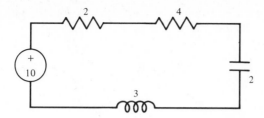

FIGURE 3—4. Network of Fig. 3—2 in the initial condition form.

10-V charge on the 2-F capacitor since the inductor can only support a voltage drop if there is a changing current, and therefore has none. So $v_{2F}(0) = 10$ V, positive on the top. Combining everything gives the result

$$4i(t) + 3\frac{di}{dt} + \frac{1}{2}\int_0^t i(\tau)\,d\tau = -5$$

The signs were determined by the clockwise direction of the current and the fact that the voltage drops across capacitor, inductor, and resistor are positive on the side of the component into which the current enters. ∎

Before considering how to solve equations such as the one that resulted in Example 3-1, a similar process for finding network equations using KCL will be introduced.

KCL. The use of KCL to derive an equation for networks with capacitors and inductors follows along the same lines as that using KVL. Use of this law does, however, show some properties of inductors that will be important to later studies. The network of Fig. 3–5 has a current source, resistor, capacitor, and inductor in a parallel arrangement, which makes application of KCL appropriate. There are two nodes, as identified in the figure. Each element has an associated current and KCL states that the sum of these currents will be zero at a node.

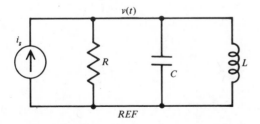

FIGURE 3—5. Network with a capacitor and inductor for illustrating the application of KCL.

Both nodes will give the same equation, so there is no need to write an equation for each node. The KCL equation is

$$i_s + i_R + i_C + i_L = 0 \tag{3-9}$$

where

$$i_s = \text{given current source}$$

$$i_R, i_C, i_L = \text{currents through resistor, capacitor, and inductor}$$

Equation (3-9) is one equation in three unknowns, the currents through the components. In order to get one equation in one unknown, as is necessary to solve the equation, the $i(t)$ versus $v(t)$ relations of each component will be used. Each component has a common voltage across it, labeled as $v(t)$ in Fig. 3–5.

For the resistor, Ohm's law shows that

$$i_R = \frac{v(t)}{R} \tag{3-10}$$

where the direction of the current will be from higher potential to lower potential. Thus, if $v(t)$ is assumed to be aboveground potential, the current will be from the node to ground (i.e., leaving the node).

For the capacitor, Eq. (1-6) will provide the necessary relationship:

$$i_C = C \frac{dv}{dt} \tag{3-11}$$

where again the current is assumed to flow from the node at $v(t)$ to the reference.

Equation (1-8) provides a relationship between the voltage and derivative of the current. But what is needed here is the reverse relationship. It will therefore be necessary to integrate Eq. (1-8) to find the current in terms of the voltage. By comparison with the integration procedure carried out in the preceding section, you can see that the result will be

$$i_L = i_L(t_0) + \frac{1}{L} \int_{t_0}^{t} v(\tau) \, d\tau \tag{3-12}$$

where

$$t_0 = \text{time origin of the problem}$$

$$t = \text{time at which the current is determined}$$

$i_L(t_0)$ = initial condition on the inductor as the current flowing
at the time origin, t_0

τ = variable of integration

Equation (3-12) shows that when considering the current through an inductor, care must be taken to include any initial current flowing through the component at the time origin of the problem. In the case of an inductor the current cannot change instantaneously. We say that the current through an inductor is *continuous*.

This gives another important rule when solving networks with capacitors and inductors:

3. Whenever inductors are present in a network the value of the current through the inductor at the time origin must be determined and included in the equations. This is called the initial condition of the inductor.

A combination of the relation derived for Fig. 3–5, Eq. (3-9), and the *i–v* relations, Eqs. (3-10), (3-11), and (3-12), will now provide the required one equation in one unknown necessary to solve the problem:

$$i_s - \frac{v(t)}{R} - C\frac{dv}{dt} - i_L(0) - \frac{1}{L}\int_0^t v(\tau)\,d\tau = 0 \qquad (3\text{-}13)$$

As in the case of KVL, the use of KCL to set up the equation of a network with capacitors and inductors results in an integrodifferential equation. The following example illustrates the development of such an equation in a specific problem.

EXAMPLE 3-2
In the network of Fig. 3–6 the switch closes at $t = 0$. Derive the integrodifferential equation for the voltage of this network using KCL.

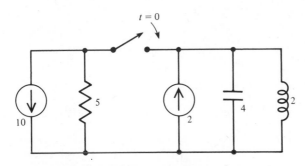

FIGURE 3–6. Network for Example 3-2.

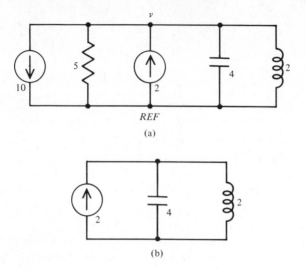

(a)

(b)

FIGURE 3–7. In (a) the actual network to be solved is shown. In (b) the initial condition is illustrated.

SOLUTION

When the switch has closed, the network has the form shown in Fig. 3–7a, where the nodes have been identified, with one selected as the reference. Application of KCL gives the equation

$$-10 - i_{5\Omega} + 2 - i_{4F} - i_{2H} = 0$$

The polarities of these terms are determined by assuming that the component currents flow from the voltage node to the reference, and the source directions are, of course, given. The resistor current is given from Eq. (3-10), $i_R = v(t)/5 = 0.2v(t)$. The capacitor current is given from Eq. (3-11) as, $i_C = 4(dv/dt)$. Inductor current is found from Eq. (3-12), but you can see that the initial condition of the inductor must be found first. This can be done by considering the network just before the switch closes, as given in Fig. 3–7b. Since the inductor is a dc short circuit it is clear that all of the 2-A current source will go through this component, so $i_L(0) = 2$, directed from voltage node to reference. If this is substituted into Eq. (3-12) and everything is combined, the result is

$$0.2v(t) + 4\frac{dv}{dt} + 0.5 \int_0^t v(\tau)\, d\tau = -10 \qquad \blacksquare$$

Current–Voltage Summary. Table 3–1 presents a summary of the $i(t)$ versus $v(t)$ relationships for resistors, capacitors, and inductors. These relations are used in combination with KVL and KCL to develop equations that will solve

TABLE 3–1

Resistor	$i(t) = \dfrac{v(t)}{R}$	$v(t) = Ri(t)$
Capacitor	$i(t) = C\dfrac{dv}{dt}$	$v(t) = v(0) + \dfrac{1}{C}\displaystyle\int_0^t i(\tau)\,d\tau$
Inductor	$i(t) = i(0) + \dfrac{1}{L}\displaystyle\int_0^t v(\tau)\,d\tau$	$v(t) = L\dfrac{di}{dt}$

any given network. So far, only single loop or node networks have been considered. In the next section it is shown that multiple loop and node networks can be developed in the same manner as has been presented. The result will be simultaneous sets of integrodifferential equations. As the entries in Table 3–1 show, the concept of time origin, t_0, and *initial conditions* appear explicitly in the current–voltage relationships of capacitors and inductors.

Relation to Reactance. The relations presented in Table 3–1 are the most general way of describing the variation of current through and voltage across resistors, capacitors, and inductors. Some may ask: "Why not just use the *reactance* of capacitors and inductors instead of integral and derivative relations?" The reason is that reactance is just one *special case* of the equations given in Table 3–1, valid only when the time dependence of the problem is purely sinusoidal. The relations of Table 3–1 are valid for any time dependence—including sinusoidal, but also square waves, triangular waves, exponential decay—anything.

To be specific, consider an inductor, L, which carries a current given by $i(t) = I_0 \sin(\omega t)$. From Table 3–1 the voltage across the inductor will be

$$v_L(t) = L\frac{di}{dt} = L\frac{d[I_0 \sin(\omega t)]}{dt}$$

$$v_L(t) = \omega L I_0 \cos(\omega t)$$

or

$$v_L(t) = \omega L I_0 \sin\left(\omega t + \frac{\pi}{2}\right)$$

So the idea of reactance for an inductor being $X_L = \omega L$ with a phase shift of $\pi/2$ radians (90°) comes from an application of the general differential relation to a sinusoidal problem.

3-1.2 Initial Conditions

One of the most important outcomes of the previous sections has been the realization that the initial conditions play a very important role in the solution of a network. The following three rules were established for solving problems with capacitors and inductors:

1. The time origin of the network must be established.
2. The initial voltage on all capacitors must be determined.
3. The initial current through all inductors must be determined.

In the final analysis the last two ideas are the result of the conservation of energy in the network, a most fundamental concept. The initial voltage on a capacitor and current through an inductor represent energy stored in these devices which will affect the future behavior of the network. The fact that the initial capacitor voltage and inductor current must be included, as in the relations of Table 3–1, is a statement of the fact that these quantities are *continuous* across the time origin of the problem.

Continuous and Discontinuous Variables.　Let us consider a network problem where, for convenience, the time origin is at $t_0 = 0$. Thus, at $t = 0$, a switch closes or opens or something, such that the problem of interest is started, and a solution is desired from that moment on. A *continuous variable* will be one whose value is the same *instantaneously* before $t = 0$ and *instantaneously* after $t = 0$. A *discontinuous variable* is one for which this is not true. Analytically, this can be described by defining $t = 0^-$ to be an infinitesimal moment before $t = 0$ and $t = 0^+$ to be an infinitesimal moment after $t = 0$. Thus, if $f(t)$ is a continuous variable, $f(0^-) = f(0^+)$; and if $g(t)$ is *not* a continuous variable, $g(0^-) \neq g(0^+)$. You may think that all network variables are continuous, but in fact most are not!

The voltages across capacitors and currents through inductors are continuous network variables: $v_C(0^-) = v_C(0^+)$ and $i_L(0^-) = i_L(0^+)$. The rest of the network variables are not necessarily continuous, although in any given problem they may be by coincidence. It is interesting that the current through capacitors and voltage across inductors are generally not continuous.

Figure 3–8a shows a simple network for which a switch moves from position a to b at $t = 0$. From elementary considerations you can see that the capacitor must have a 10-V charge with the switch in position a, since no dc current can flow through the capacitor. Thus the initial voltage is 10 V and the initial current is zero; $v_C(0^-) = 10$ and $i_C(0^-) = 0$. At $t = 0$ the switch moves to position b, which places 10 V across the 5-Ω resistor. By Ohm's Law, the instant the switch closes the current through the resistor must be $i_R = 10/5 = 2$ A. This means that $i_C(0^+) = 2$ A and, of course, $v_C(0^+) = 10$ V. Thus the current was *discontinuous* across $t = 0$. This is shown by the graphs of current and voltage versus time in Fig. 3–8b.

Initial Steady State.　In many cases a network will be in an initial state which is purely dc; that is, no time variation of currents or voltages is occurring.

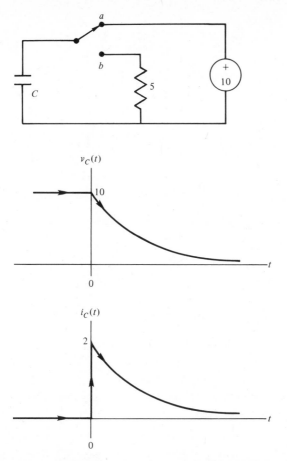

FIGURE 3–8. When the switch moves from position *a* to position *b* the capacitor voltage will be continuous but the current will be discontinuous.

For this type of problem the initial conditions can be found by using resistive network analysis to solve the network in the initial state, $t = 0^-$. The general procedure is as follows:

1. Replace all capacitors by open circuits, noting the location of the capacitor terminals.
2. Replace all inductors by short circuits, noting the location of the inductor terminals.
3. Solve the resulting resistive network, using mesh or nodal analysis, for example. The voltage across the capacitor terminal locations will be the capacitor initial conditions. The currents through the inductor locations will be the inductor initial conditions.

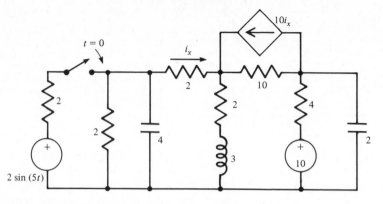

FIGURE 3–9. Network for Examples 3-3 and 3-5.

EXAMPLE 3-3

Find the initial conditions of the network given in Fig. 3–9. The switch closes at $t = 0$.

SOLUTION

Although there is a sinusoidal source, in the $t = 0^-$ state it is not part of the network; thus there is an initial steady state. The network in Fig. 3–10 results from replacing all capacitors by open circuits and inductors by short circuits. The result is a resistive network. The ICIS has been transformed to an ICVS in anticipation of using mesh analysis to solve the problem. The required initial conditions will be

$$v_{4F}(0^-) = -2i_1$$

$$i_{3H}(0^-) = i_1 - i_2$$

$$v_{2F}(0^-) = 4i_2 + 10$$

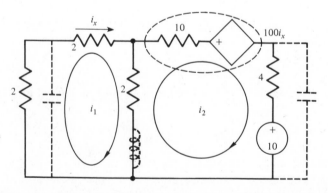

FIGURE 3–10. Network of Fig. 3–9 in the initial state and set up to find the initial conditions by mesh analysis.

CAPACITORS AND INDUCTORS

Using mesh analysis and the loop currents shown, the matrix equation is

$$\begin{pmatrix} 6 & -2 \\ -2 & 16 \end{pmatrix} \begin{pmatrix} i_1 \\ i_2 \end{pmatrix} = \begin{pmatrix} 0 \\ -100 i_x - 10 \end{pmatrix}$$

From the network it is clear that $i_x = i_1$, so the matrix equation becomes

$$\begin{pmatrix} 6 & -2 \\ 98 & 16 \end{pmatrix} \begin{pmatrix} i_1 \\ i_2 \end{pmatrix} = \begin{pmatrix} 0 \\ -10 \end{pmatrix}$$

Solving by Cramer's rule gives the following currents: $i_1 = -0.068$ A, $i_2 = -0.205$ A. The initial conditions are found from the relations above as

$$v_{4F}(0^-) = +0.136 \text{ V}$$

$$i_{3H}(0^-) = 0.137 \text{ A}$$

$$v_{2F}(0^-) = 9.18 \text{ V} \qquad \blacksquare$$

Initial Dynamic State. In some cases the time origin of one problem will be the solution to another, time-dependent problem at some terminal time. In these cases the first problem is solved to find the initial conditions of the second problem. In Fig. 3–11 a situation is given which describes this type of problem. In position 1 the network is in the initial state. Steady-state analysis would determine that $v_C(0) = 0$ and $i_L(0) = 0$. These are the initial conditions for the problem when the switch moves to position 2. Suppose that after 2 s the switch is moved to position 3. Then the *initial* conditions for this problem (in position 3) would be the values of capacitor voltage and inductor current calculated in position 2 at $t = 2$ s.

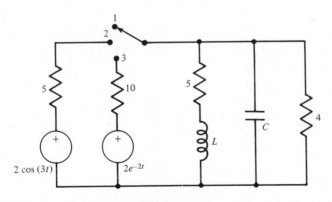

FIGURE 3–11. Network for Example 3-4.

GENERAL NETWORK EQUATIONS

The preceding section showed how the presence of capacitors and inductors in a network results in integrodifferential equations to be solved to find the network variables. In order to emphasize the properties of capacitors and inductors, previous examples involved only single loop and two node problems. In this section procedures for setting up equations for more general networks will be presented, as well as the introduction of transformers in the networks. At this point the objective is merely to see how the equations can be derived and not, yet, how these equations can be solved for the network variables themselves.

3-2.1 KVL

In applying KVL to a general network problem the following procedures can be followed to derive equations for the network variables.

1. Determine the time origin and all initial conditions.
2. Transform all current sources to voltage sources.
3. Assign clockwise currents to each elementary loop. These are the network variables.
4. Use KVL and the i–v relationships of Table 3–1 to construct a set of simultaneous integrodifferential equations for the network variables.

Note: Steps 2 and 3 are not *required* but can make derivation of the equations follow a formal structure, which is usually easier.

EXAMPLE 3-4

Find the equations to solve the network of Fig. 3–11 for $t > 0$ if the switch moves from position 1 to position 2 at $t = 0$. Use $C = 2$ F and $L = 3$ H.

SOLUTION

First note that since the switch never moves to position 3, the resistor and source connected to this position can be eliminated from the network. Now, position 1 establishes the initial conditions. It is clear that $v_{2F}(0) = 0$ and also $i_{3H}(0) = 0$ since there are no sources connected in this initial state. Figure 3–12 shows the network to be solved (position 2) with clockwise currents assigned to each loop. By KVL applied to each loop, and using the relations of Table 3–1, the following equations result.
 Loop 1:

$$2 \cos (3t) - 5i_1 - 5(i_1 - i_2) - 3 \frac{d(i_1 - i_2)}{dt} = 0$$

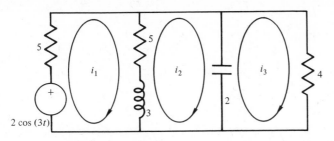

FIGURE 3–12. This is the network to be solved in Example 3-4, set up with loop currents for application of KVL.

Loop 2:

$$5(i_2 - i_1) + \tfrac{1}{2} \int_0^t (i_2 - i_3)\, d\tau + 3\,\frac{d(i_2 - i_1)}{dt} = 0$$

Loop 3:

$$\tfrac{1}{2} \int_0^t (i_3 - i_2)\, d\tau + 4i_3 = 0$$

This set of equations can be simplified by noting that the derivatives and integrals of differences between variables can be broken into the differences of the derivatives and integrals. When this is done, as well as some rearrangement of the equations, the following set of integrodifferential equations result:

$$10i_1 + 3\,\frac{di_1}{dt} - 5i_2 - 3\,\frac{di_3}{dt} = 2\cos(3t)$$

$$-5i_1 - 3\,\frac{di_1}{dt} + 5i_2 + 3\,\frac{di_2}{dt} + \tfrac{1}{2}\int_0^t i_2\, d\tau - \tfrac{1}{2}\int_0^t i_3\, d\tau = 0$$

$$-\tfrac{1}{2}\int_0^t i_2\, d\tau + 4i_3 + \tfrac{1}{2}\int_0^t i_3\, d\tau = 0 \qquad ■$$

3-2.2 KCL

To apply KCL to a network in order to derive the equations for the network variables, the following procedure can be followed:

1. Determine the time origin and all the initial conditions.
2. Transform all the sources to current sources. This step is not necessary but makes the formal procedure easier to follow.
3. Identify all nodes, select one as the reference, and assign to the rest unknown voltages. These are the network variables.
4. Use KCL and the relations of Table 3–1 to set up the simultaneous equations for the node voltages.

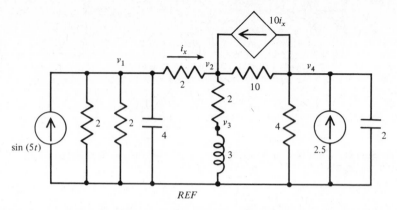

FIGURE 3–13. Network of Fig. 3–9 set up for finding the equations using KCL.

EXAMPLE 3-5

The switch in the network of Fig. 3–9 closes at $t = 0$. Construct the set of simultaneous equations that will solve this network.

SOLUTION

The initial conditions of this problem were found in Example 3-3. Only the initial current through the inductor shows up at this stage and this was found to be $i_{3H}(0) = 0.137$ A. The sinusoidal source in series with the 2-Ω resistor is transformed into a current source and the network is redrawn in Fig. 3–13 with node identification. Using KCL and the relations of Table 3–1, the node equations can be written as

Node 1:

$$\sin (5t) - \frac{1}{2}v_1 - \frac{1}{2}v_1 - 4\frac{dv_1}{dt} - \frac{v_1 - v_2}{2} = 0$$

Node 2:

$$10i_x + \frac{v_1 - v_2}{2} - \frac{v_2 - v_4}{10} - \frac{v_2 - v_3}{2} = 0$$

Node 3:

$$\frac{v_2 - v_3}{2} - \frac{1}{3}\int_0^t v_3 \, d\tau - 0.137 = 0$$

Node 4:

$$-10i_x + \frac{v_2 - v_4}{10} - \frac{1}{4} + 2.5 - 2\frac{dv_4}{dt} = 0$$

CAPACITORS AND INDUCTORS

These equations need to be rearranged, but also the dependency expressed as i_x must be expressed in terms of the network variables. This current is the total current through a 2-Ω resistor connected between nodes 1 and 2. Thus, by Ohm's law,

$$i_x = \frac{v_1 - v_2}{2}$$

Then $10i_x = 5v_1 - 5v_2$. Inserting this into the equations above and simplifying gives the final set of integrodifferential equations:

$$1.5v_1 + 4\frac{dv_1}{dt} - 0.5v_2 = \sin(5t)$$

$$-5.5v_1 + 6.1v_2 - 0.5v_3 - 0.1v_4 = 0$$

$$-0.5v_2 + 0.5v_3 + \frac{1}{3}\int_0^t v_3\, d\tau = -0.137$$

$$5v_1 - 5.1v_2 + 0.35v_4 + 2\frac{dv_4}{dt} = 2.5$$

3-2.3 Transformers

Transformers are included in the equations for a network using the relations given in Chapter 1 between the voltages across transformer windings and the currents through the windings. There is an important condition on the use of Eqs. (1-10) and (1-11) to represent the voltage drops across the transformer windings. These equations are written in accordance with the voltage polarity and current directions assigned to the transformer in Fig. 1–6. Thus, when writing loop equations, care must be taken to keep agreement between the network variables and the conventions of the transformer equations.

In summary, notice in Fig. 1–6 that the voltages are assumed positive on the top of both windings and that the currents are assumed to enter the top of both windings. Under these conditions Eqs. (1-10) and (1-11) are valid. If different *assigned* voltage polarities or current directions are used, appropriate sign changes must be made in the equations.

A network with a transformer may be approached by using the transformer in the network directly or by using a model, such as the T inductor model presented in Fig. 1–8 and defined by Eqs. (1-13) and (1-14). If the transformer is used directly, KVL is easy to apply since the voltages of each winding are specified in terms of current. When KCL is to be used to solve the network, it is easier to use the T model (if such a model fits the particular problem) since the current equations for these components can be written from the relations of Table 3–1. The following examples illustrate how the network equations can be derived with transformers in the network.

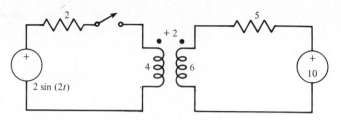

FIGURE 3–14. Network for Example 3-6.

EXAMPLE 3-6

The network of Fig. 3-14 has the switch closed at $t = 0$. Find the network equations using KVL.

SOLUTION

The first problem will be to establish the initial conditions. On the left there is clearly no current since the switch is open at $t = 0^-$. On the right, however, there is a steady current since the transformer winding is a dc short circuit. The current through the transformer is simply $i_{6H}(0^-) = 10/5 = 2$ A. Figure 3–15 shows the network set up for KVL. The voltage drops across the transformer have been labeled v_p and v_s with polarities as defined in Fig. 1–6. The normal clockwise loop currents have also been indicated. Notice that the secondary current, i_2, has the *opposite* direction to that defined in Fig. 1–6. Therefore, a negative sign will appear in all terms that contain i_2 and Eqs. (1–10) and (1–11) become

$$v_p = 4\frac{di_1}{dt} - 2\frac{di_2}{dt}$$

$$v_s = -6\frac{di_2}{dt} + 2\frac{di_1}{dt}$$

Using these relations, KVL can be applied to the network to write equations for the two loops, summing the voltage sources and drops to zero as usual. After

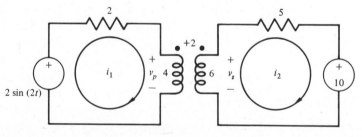

FIGURE 3–15. Network of Fig. 3–14 set up for KVL.

doing this and rearranging, the result is the following set of differential equations:

$$2i_1 + 4\frac{di_1}{dt} - 2\frac{di_2}{dt} = 2 \sin (2t)$$

$$-2\frac{di_1}{dt} + 5i_2 + 6\frac{di_2}{dt} = -10$$

EXAMPLE 3-7
Use the T model of a transformer to set up the equations to solve the network given in Fig. 3–14.

SOLUTION
The T model is defined by the schematic of Fig. 1–8 and by Eqs. (1-13) or Eq. (1-14). In this case the effect of the mutal inductance is positive, so Eqs. (1-13) are used. The values of the three inductances are

$$L_A = L_1 - M = 4 - 2 = 2 \text{ H}$$

$$L_B = L_2 - M = 6 - 2 = 4 \text{ H}$$

$$L_C = M = 2 \text{ H}$$

Using these the network is redrawn, with the switch closed, as shown in Fig. 3–16, with loop currents assigned. Now the problem is simply treated as one with three inductors. Applying the KVL approach outlined above results in the following two equations:

$$2 \sin (2t) - 2i_1 - 2\frac{di_1}{dt} - 2\frac{d(i_1 - i_2)}{dt} = 0$$

$$4\frac{di_2}{dt} + 5i_2 + 10 + 2\frac{d(i_2 - i_1)}{dt} = 0$$

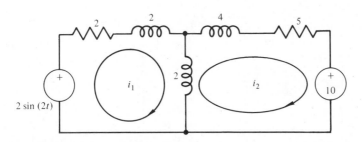

FIGURE 3–16 Network of Fig. 3–14 using a T model for the transformer as specified in Example 3-7.

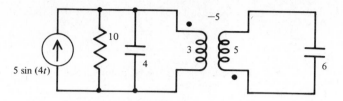

FIGURE 3–17. Network for Example 3-8.

When these equations are simplified and rearranged, the result is the same as that given in Example 3-6—as, of course, it must be. ∎

EXAMPLE 3-8

Given the network of Fig. 3–17, derive the equations that will solve this network using KCL. Assume that the time origin is $t = 0$ and that the initial conditions are all zero.

SOLUTION

For this problem the T model of the transformer will be used. The connection is such that the effect of the mutual inductance is negative, so that Eqs. (1-14) are used to find the inductors of the model,

$$L_A = L_1 + |M| = 3 + 5 = \;\; 8 \text{ H}$$

$$L_B = L_2 + |M| = 5 + 5 = 10 \text{ H}$$

$$L_C = -|M| = -5 \text{ H}$$

The network has been redrawn in Fig. 3–18 with the T model inserted and the nodes defined and labeled. Note that the node *within* the model must be used, even though this node does not physically exist in the network. Applying KCL and the relations of Table 3–1 provides the following equations, after rearrangement:

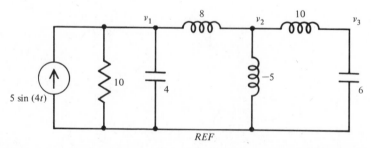

FIGURE 3–18. Network of Fig. 3–17 with a T model of the transformer and set up for KCL.

CAPACITORS AND INDUCTORS

$$0.1v_1 + 4\frac{dv_1}{dt} + 0.125 \int_0^t v_1 \, d\tau - 0.125 \int_0^t v_2 \, d\tau = 5 \sin{(4t)}$$

$$-0.125 \int_0^t v_1 \, d\tau + 0.025 \int_0^t v_2 \, d\tau - 0.1 \int_0^t v_3 \, d\tau = 0$$

$$-0.1 \int_0^t v_2 \, d\tau + 0.1 \int_0^t v_3 \, d\tau + 6\frac{dv_3}{dt} = 0 \qquad \blacksquare$$

3-2.4 Solving General Network Equations

The previous sections of this chapter have shown how the equations that will solve a general linear network can be derived from KVL and KCL together with the i–v relations for resistors, capacitors, inductors, and transformers. The most general resulting equations will involve both integrals and derivatives of the network variables. The next topic to be considered is how such equations can be solved for these network variables.

Differential Equations.   The traditional method of solving integrodifferential equations is by converting them to entirely differential equations. Methods for solving differential equations have been studied for many years, so the approaches are well established, especially for the types that appear from electrical networks. Although this method often gives insights into network behavior, it is not the most efficient method of solving the types of equations that result from networks, particularly simultaneous sets of differential equations. The remainder of this chapter deals with the techniques of solving differential equations by this classical approach, although limited to problems that result in single equations.

Laplace Transforms.   In Chapter 4 another method is presented for solving differential equations, or rather, integrodifferential equations. This method does not require that the equations be converted into differential equations and can handle simultaneous sets of such equations. Eventually, application of Laplace techniques allows a return to the highly efficient methods of matrix representation of the simultaneous set of equations and even construction of the matrices by inspection from the network (almost).

3-3

DIFFERENTIAL EQUATIONS

A differential equation is one in which the unknown function appears with its derivatives. The equation may contain coefficients of the function and its derivatives and other terms that are constant or functions of the independent variable.

The most general form of a linear nth-order differential equation can be written as

$$c_n(t) \frac{d^n f}{dt^n} + c_{n-1}(t) \frac{d^{n-1} f}{dt^{n-1}} + \cdots + c_1(t) \frac{df}{dt} + c_0(t)f(t) = F(t) \quad (3\text{-}14)$$

where

$$t = \text{independent variable (time in network problems)}$$
$$c_n(t), c_{n-1}(t), \ldots, c_0(t) = \text{coefficients}$$
$$f(t) = \text{unknown function}$$
$$F(t) = \text{given forcing function}$$

This is called nth order because the highest derivative is an nth-order derivative. It is called linear because the unknown function never appears with a power greater or less than 1. Studies of differential equations such as the one in Eq. (3-14) have occupied science and mathematics for many years since many physical problems of interest end up requiring the solution of a differential equation. You have already seen this in the electrical sciences with networks. Fortunately, the types of differential equations that result from network analysis are of a much simpler form than the general type given above.

3-3.1 Differential Equations of Networks

In nearly all of the problems that arise in the linear networks of electricity, only two specific types of differential equations are found.

1. Linear first-order differential equations with constant coefficients:

$$c_1 \frac{df}{dt} + c_0 f(t) = F(t) \quad (3\text{-}15)$$

2. Linear second-order differential equations with constant coefficients:

$$c_2 \frac{d^2 f}{dt^2} + c_1 \frac{df}{dt} + c_0 f(t) = F(t) \quad (3\text{-}16)$$

In both cases the coefficients are constants; that is, the coefficients do not depend on the independent variable, time.

Neither Eq. (3-15) nor Eq. (3-16) contain any integrals. Yet the equations that resulted from network analysis often contain integrals. This difference is resolved by *differentiating* the equations that result from network analysis when they contain integrals. For example, consider Eq. (3-8), which was a result of applying KVL to a simple network. You can see that this equation contains the integral of the current as well as the first derivative. If this equation is differentiated, the result, after rearrangement, is

$$L\frac{d^2i}{dt^2} + R\frac{di}{dt} + \frac{1}{C}i(t) = \frac{dV}{dt} \qquad (3\text{-}17)$$

Now, Eq. (3-17) is a linear second-order differential equation with constant coefficients. The forcing function is $F(t) = dV/dt$. All of the equations that result from linear networks will result in one of the two forms, Eq. (3-15) or (3-16). The next sections will present the methods by which these equations can be solved for the unknown function.

Unknown Constants. The solution of a differential equation will always contain as many unknown constants as the order of the equation, two for second order and one for first order. The values of these constants are found by using the *initial conditions* which have been established at the beginning of the problem. This is done by deducing what the value of the unknown function and/or its derivative is at the time origin of the problem, employing the initial conditions to perform this deduction.

Suppose, as an example, that the solution of some problem turns out to be a current, $i(t) = Ke^{-3t}$, where K is the unknown constant. Perhaps from the initial conditions we find that the value of the current at the time origin is $i(0) = 10$ A. Then by requiring the differential equation solution to match this fact, the constant is determined, $i(0) = Ke^0 = K = 10$. Thus the complete solution is $i(t) = 10e^{-3t}$.

3-3.2 First-Order Differential Equations

First-order differential equations of the type given by Eq. (3-15) are frequently found in network analysis. They arise from problems that contain either capacitors only or inductors only, but not both. To define the solution of this type of equation let us first factor c_1 from Eq. (3-15) and write the result in the form

$$\frac{df}{dt} + af(t) = H(t) \qquad (3\text{-}18)$$

where

$$a = c_0/c_1 \text{ is a constant}$$

$$H(t) = F(t)/c_1 \text{ is the forcing function}$$

It is possible to write down a general solution for Eq. (3-18) which can be evaluated for any forcing function, $H(t)$. This solution is

$$f(t) = Ke^{-at} + e^{-at} \int_{t_0}^{t} H(\tau)e^{a\tau} \, d\tau \qquad (3\text{-}19)$$

where

$$K = \text{unknown constant}$$

$$\tau = \text{variable of integration}$$

$$t_0 = \text{the time origin of the problem}$$

The first term in Eq. (3-19) is known as the *homogeneous* solution. This term comes from solving Eq. (3-18) with $H(t) = 0$. Since this solution results with no forcing function, it is considered the *natural* response of the network. Notice that this term contains an undetermined constant, K. The second term is known as the *particular* solution of the differential equation and results from the specific form of the forcing function, $H(t)$.

The following sections consider first the homogeneous solution and then several common forms for the forcing function and the resulting form of the complete solution, as found from Eq. (3-19).

Homogeneous. The term *homogeneous* is used to describe those cases where the forcing function is zero [i.e., $H(t) = 0$]. In these cases the form of the differential equation is simply

$$\frac{df}{dt} + af(t) = 0 \qquad (3\text{-}20)$$

The solution is easily found from Eq. (3-19) by using $H(t) = 0$,

$$f(t) = Ke^{-at} \qquad (3\text{-}21)$$

All that remains is to determine the value of K. One of the easiest ways to do this is to find the value of $f(0)$, since you can see from Eq. (3-21) that $f(0) = K$.

EXAMPLE 3-9
Find the current in the network of Fig. 3–19. The switch closes at $t = 0$.

SOLUTION
First the initial conditions must be established. Clearly, there is no charge on the capacitor initially since the source is isolated by the switch for $t < 0$, so

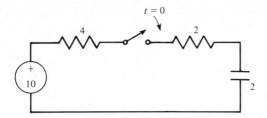

FIGURE 3–19. Network for Example 3-9.

$v_{2F}(0) = 0$. Now KVL is applied to the network together with the relations of Table 3–1. The resulting equation is

$$10 - 4i(t) - 2i(t) - \tfrac{1}{2} \int_0^t i(\tau)\, d\tau = 0$$

This equation is now differentiated and rearranged, giving the differential equation

$$\frac{di}{dt} + \frac{1}{12} i(t) = 0$$

This equation is a homogeneous first-order differential equation of the type given by Eq. (3-20) with $a = \tfrac{1}{12}$. The solution is thus

$$i(t) = Ke^{-t/12}$$

The constant K is found from the initial conditions as follows. When the switch closes there will be a net of 10 V across the 4- and 2-Ω resistors since the capacitor is uncharged. Thus the initial current must be

$$i(0^+) = \frac{10}{4 + 2} = 1.67 \text{ A}$$

This is to be matched with the solution above,

$$i(0) = Ke^0 = K = 1.67$$

So the solution is

$$i(t) = 1.67e^{-t/12} \qquad \blacksquare$$

Constant Forcing Function. In many cases the forcing function is a constant, $H(t) = H_0$. The form of the differential equation then becomes

$$\frac{df}{dt} + af(t) = H_0 \tag{3-22}$$

and the solution is found from Eq. (3-19) by integration:

$$f(t) = Ke^{-at} + e^{-at}H_0 \int_{t_0}^{t} e^{a\tau} \, d\tau$$

or

$$f(t) = Ke^{-at} + \frac{H_0}{a}[1 - e^{-a(t-t_0)}] \tag{3-23}$$

EXAMPLE 3-10

Find the voltage and current of the capacitor in the network of Fig. 3-20 after the switch moves from position a to position b at $t = 0$. The switch had been in position a for a long time.

SOLUTION

The initial conditions are established with the switch in position a. Since a capacitor is a dc open circuit all 4 A flows through the parallel 2- and 6-Ω resistors. From this the voltage on the capacitor is found as the voltage dropped across this parallel combination, $v_C(0) = (4)(12/8) = 6$ V, positive on the top of the capacitor. Figure 3-21 shows the network to be solved with the switch in position b. Using KCL and the i–v relations results in the following equation for the node voltage:

$$4 - \frac{v(t)}{1} - \frac{v(t)}{6} - 2\frac{dv}{dt} = 0$$

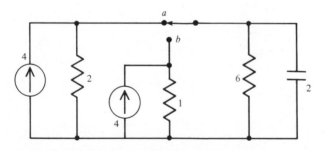

FIGURE 3-20.   Network for Example 3-10.

CAPACITORS AND INDUCTORS

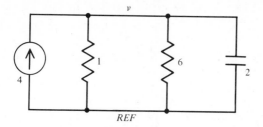

FIGURE 3–21. Network to be solved in Example 3-10.

Simplifying, we obtain

$$\frac{dv}{dt} + \frac{7}{12} v(t) = 2$$

This equation is of the form given by Eq. (3-22), where the forcing function is a constant. The solution, with the time origin equal to zero, is then given by Eq. (3-23) as

$$v(t) = Ke^{-7t/12} + \tfrac{24}{7} (1 - e^{-7t/12})$$

The value of the unknown constant is found directly from the initial condition, since the node voltage and capacitor voltage are the same. Thus

$$v(0) = 6 = K + \tfrac{24}{7} (1 - 1) = K$$

So $K = 6$ and the complete solution is

$$v(t) = 6e^{-7t/12} + \tfrac{24}{7} (1 - e^{-7t/12})$$

or

$$v(t) = \tfrac{24}{7} + \tfrac{18}{7} e^{-7t/12}$$

The current through the capacitor is found by differentiating the voltage equation,

$$i_C(t) = C \frac{dv}{dt} = 4\left(\frac{d}{dt}\right) \left[\frac{24}{7} + \frac{18}{7} e^{-7t/12} \right]$$

$$i_C(t) = -6e^{-7t/12}$$

The current and voltage are plotted in Fig. 3–22. Note that the voltage is continuous across the time origin, whereas the current is not. ∎

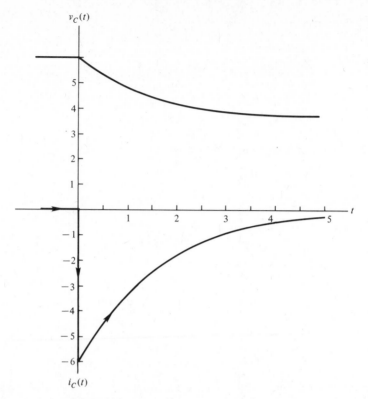

FIGURE 3–22. Plot of capacitor current and voltage for the network in Fig. 3–20.

General Forcing Function. In many cases $H(t)$ will be the signal of a generator or other source with some specified time dependence. In these cases the complete solution as given by Eq. (3-19) must be solved for the particular $H(t)$ of the source. The following example illustrates this for one type of signal source and also the concept of *dynamic initial conditions*.

EXAMPLE 3-11

In the network of Fig. 3–23, the switch moves to position 1 at $t = 0$ and then to position 2 at $t = 2$ s. Find the current for each switch position and plot.

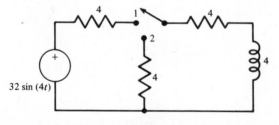

FIGURE 3–23. Network for Example 3-11.

CAPACITORS AND INDUCTORS

SOLUTION

It is clear that in the initial state ($t = 0^-$) no current can flow and hence $i_{4H}(0) = 0$ A. When the switch is moved to position 1, KVL can be easily used to show that the resulting differential equation is

$$\frac{di}{dt} + 2i(t) = 8 \sin (4t)$$

The solution for this equation can be written down formally from Eq. (3-19), in terms of an integral,

$$i(t) = Ke^{-2t} + e^{-2t} \int_0^t 8 \sin (4\tau)e^{2\tau} \, d\tau$$

The integral in this equation can either be evaluated directly or looked up in a set of integral tables. Choosing the latter, the integral form is found from a table of integrals as

$$\int e^{ay} \sin (y) \, dy = \frac{e^{ay}}{a^2 + 1} [a \sin (y) - \cos (y)]$$

comparison to the integral in the expression for $i(t)$ shows that $y = 4\tau$ and $a = \frac{1}{2}$; thus

$$\frac{1}{4} \int_0^t e^{1/2(4\tau)} \sin (4\tau) \, d4\tau = \frac{1}{4} \frac{e^{2\tau}}{4\frac{1}{4} + 1} \left[\frac{1}{2} \sin (4\tau) - \cos (4\tau)] \right]_0^t$$

When this is simplified by substitution of the upper and lower limits and the result is placed in the solution for $i(t)$ found above, the solution becomes

$$i(t) = Ke^{-2t} + \tfrac{4}{5} \sin (4t) - \tfrac{8}{5} \cos (4t) + \tfrac{8}{5}e^{-2t}$$

The initial condition on the inductor shows that the current is zero at $t = 0$. When this is substituted into the solution, the constant K can be found:

$$i(0) = 0 = K + 0 - \tfrac{32}{5} + \tfrac{32}{5} = K \qquad \text{or} \qquad K = 0$$

The final solution for the current in switch position 1 is thus given by

$$i(t) = \tfrac{8}{5}e^{-2t} + \tfrac{4}{5} \sin (4t) - \tfrac{8}{5} \cos (4t)$$

Well, this solves half the problem. Now, at $t = 2$ the switch moves to position 2. In this case the problem to be solved has the form given in Fig. 3–24. The

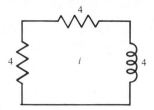

FIGURE 3–24. Network to be solved in the second part of Example 3-11, for $t > 2$.

initial condition for this problem will be the current through the inductor, from the first solution, at $t = 2$ s. This is

$$i(2) = \tfrac{8}{5}e^{-4} + \tfrac{4}{5}\sin(8) - \tfrac{8}{5}\cos(8)$$

$$i(2) = 1.05 \quad \text{(rounded)}$$

Applying KVL to the network of Fig. 3–24 easily results in the differential equation

$$\frac{di}{dt} + 2i(t) = 0$$

This is a homogeneous equation and has the solution

$$i(t) = K_2 e^{-2t}$$

where K_2 is an unknown constant. The value of the unknown constant is found by matching to the initial condition at 2 s:

$$i(2) = 1.05 = K_2 e^{-4}$$

or

$$K_2 = 1.05e^4$$

Now, this result is used to provide a complete solution in switch position 2:

$$i(t) = 1.05e^{-2(t-2)} \quad (t > 2)$$

Figure 3–25 shows a plot of the current from $t = 0$ to $t = 3$ s. Notice that the solutions are continuous across $t = 0$ and $t = 2$ since an inductor current is always continuous. ∎

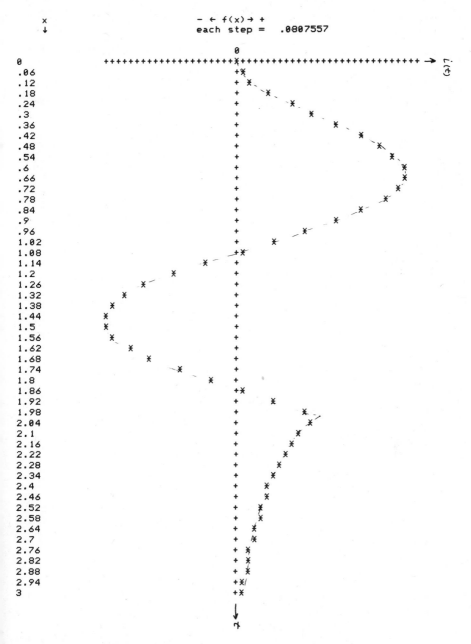

FIGURE 3—25. Plot of inductor current for Example 3-11.

3-3.3 Second-Order Differential Equations

Second-order differential equations result from networks that contain both capacitors and inductors or transformers. An examination of the results of the preceding section shows that natural response of a first-order equation is always some kind of exponential decay. The only other signals to appear came from forcing functions. Networks with capacitors and inductors, and therefore second-order equations, can give rise to exponential decay *and* oscillations (i.e., sine and cosine functions in the solution). In addition, there will now be two undetermined constants in the solutions.

The general form of the second-order differential equation can be formed from Eq. (3-16) by factoring c_2 from the expression

$$\frac{d^2f}{dt^2} + b\frac{df}{dt} + cf(t) = H(t) \tag{3-24}$$

where

$$b = c_1/c_2$$

$$c = c_0/c_2$$

$$H(t) = F(t)/c_2$$

A general solution cannot be given for the second-order differential equation, as was done for the first-order, because the *form* of the solution depends on the relative magnitudes of coefficients b and c in Eq. (3-24). It is possible to categorize the possible solutions according to the b and c coefficients *and* the form of the forcing function.

In general, the complete solution of Eq. (3-24) can be written as the sum of two parts,

$$f(t) = f_h(t) + f_p(t) \tag{3-25}$$

where

$f_h(t)$ = solution of the *homogeneous* equation defined by setting $H(t) = 0$ in Eq. (3-24)
$f_p(t)$ = solution of the *particular* equation for the specific $H(t)$ of the problem

The two solutions can be found by independent processes. The total result, obtained by adding the two solutions together, will have two undetermined constants. These constants will be found from the initial conditions of the problem by determination of the values of the complete solution and its derivative at the time origin.

Homogeneous Equation. The homogeneous second-order differential equation is

$$\frac{d^2 f_h}{d^2 t} + b \frac{df_h}{dt} + cf_h(t) = 0 \qquad (3\text{-}26)$$

The solution of this equation will have one of three forms, depending on the relative magnitude of b^2 and $4c$.

Case I. $b^2 > 4c$. In this case the solution is two exponential decay functions, so the homogeneous solution can be written

$$f_h(t) = K_1 e^{-a_1 t} + K_2 e^{-a_2 t} \qquad (3\text{-}27)$$

where

$$a_1 = \frac{b}{2} + \frac{(b^2 - 4c)^{1/2}}{2} \qquad (3\text{-}28)$$

$$a_2 = \frac{b}{2} - \frac{(b^2 - 4c)^{1/2}}{2} \qquad (3\text{-}29)$$

and K_1 and K_2 are the undetermined constants.

Case II. $b^2 = 4c$. If this condition is found to be satisfied, the solution of Eq. (3-26) will have the form

$$f_h(t) = K_1 e^{-bt/2} + K_2 t e^{-bt/2} \qquad (3\text{-}30)$$

Case III. $b^2 < 4c$. In this case the solution will have the form

$$f_h(t) = e^{-\alpha t}[K_1 \sin(\omega t) + K_2 \cos(\omega t)] \qquad (3\text{-}31)$$

where

$$\alpha = \frac{b}{2} \qquad (3\text{-}32)$$

$$\omega = \frac{(4c - b^2)^{1/2}}{2} \qquad (3\text{-}33)$$

In all three cases the undetermined constants, K_1 and K_2, will be found from the initial conditions applied to the complete solution. Cases I and II do not result in any oscillation, just exponential decay. Case III has oscillation but the amplitude of the oscillation is decaying exponentially. All of these results are therefore transient in nature. The one exception to this occurs when $b = 0$. In this case you can see that $\alpha = 0$ and thus, from Eq. (3-31), there will be a steady-state oscillation of frequency, $\omega = \sqrt{c}$. In ideal circuits this condition occurs when the resistance in a capacitor–inductor network is zero. In practical circuits this can only occur when there is a controlled source which injects energy into the network to compensate for that lost by resistance.

Undetermined Constants. The undetermined constants of the homogeneous solution are found by applying initial conditions of the network to the complete solution. In particular, the value of the equation function at zero, $f(0)$, and the value of the derivative of this function at zero, $(df/dt)_{t=0} = f'(0)$, are typically used to find these constants. The value of the function at zero is found from determination of the initial conditions of capacitors and inductors in the network. The value of the derivative at zero is usually most easily found from application of KCL or KVL to the network at $t = 0$. The examples that follow illustrate these steps.

EXAMPLE 3-12

In Example 3-1 the integrodifferential equation for the network of Fig. 3–2 was found. Express the result as a differential equation and solve for the current.

SOLUTION

The equation found in Example 3-1 was

$$4i(t) + 3\frac{di}{dt} + \tfrac{1}{2}\int_0^t i(\tau)\,d\tau = -5$$

If this equation is differentiated, a second-order differential equation will result. When this is done and the resulting equation is rearranged to be in the form of Eq. (3-24), the result is

$$\frac{d^2i}{dt^2} + \frac{4}{3}\frac{di}{dt} + \frac{1}{6}i(t) = 0$$

You can see that this equation is homogeneous from the start (i.e., there is no particular solution). This means that the solution of this homogeneous equation is the complete solution of the problem. To determine the form of the solution a comparison must be made between b^2 and $4c$. Since $b = \tfrac{4}{3}$, then $b^2 = \tfrac{16}{9}$ and $4c = 4(\tfrac{1}{6}) = \tfrac{2}{3}$, so it is clear that $b^2 > 4c$ and we have case I. The solution is found from Eq. (3-27), with a_1 and a_2 given by Eqs. (3-28) and (3-29):

$$a_1 = \frac{4}{6} + \frac{(16/9 - 2/3)^{1/2}}{2} = 1.19$$

$$a_2 = \frac{4}{6} - \frac{(16/9 - 2/3)^{1/2}}{2} = 0.14$$

Thus the solution is

$$i(t) = K_1 e^{-1.19t} + K_2 e^{-0.14t}$$

The two undetermined constants are found from the value of $i(0)$ and di/dt evaluated at $t = 0$. These can be determined from the initial conditions. From Example 3-1 the initial capacitor voltage was found to be $v_C(0) = 10$. It is clear that the inductor current is zero at $t = 0$ since there is no current at $t = 0^-$ with the switch open. Thus one of the conditions is $i(0) = 0$. The second condition can be found by consideration of the voltage across the inductor at $t = 0$, since $v_L = L(di/dt)$. After the switch is closed, application of KVL results in the equation

$$5 - v_R(0) - v_C(0) - v_L(0) = 0$$

but since $i(0) = 0$, then $v_R(0) = 0$ and we have already seen that $v_C(0) = 10$ V. Thus

$$5 - 10 - v_L(0) = 0$$

or

$$v_L(0) = L \left. \frac{di}{dt} \right]_{t=0} = 3 \left. \frac{di}{dt} \right]_{t=0} = -5$$

So the second condition is $i'(0) = -\frac{5}{3}$. The undetermined constants can now be found. Using the first initial condition, $i(0) = 0$, gives

$$i(0) = K_1 e^0 + K_2 e^0 = K_1 + K_2 = 0$$

or simply that $K_1 = -K_2$. The second condition, $i'(0) = -\frac{5}{3}$, gives

$$-5/3 = \frac{d}{dt} [K_1 e^{-1.19t} + K_2 e^{-0.14t}] \Big]_{t=0}$$

$$-5 = 3[-1.19K_1e^{-1.19t} - 0.14K_2e^{-0.14t}]\Big|_{t=0}$$

$$-5 = -3.57K_1 - 0.42K_2$$

but since $K_1 = -K_2$,

$$-5 = 3.57K_2 - 0.42K_2 = 3.15K_2$$

Thus $K_2 = -1.59$, so $K_1 = +1.59$. The solution is

$$i(t) = +1.59e^{-1.19t} - 1.59e^{-0.14t} \qquad \blacksquare$$

EXAMPLE 3-13

In Example 3-2 the integrodifferential equation for the network of Fig. 3–6 was derived. Convert this to a differential equation and find the solution.

SOLUTION

The equation derived for the network was

$$0.2v(t) + 4\frac{dv}{dt} + 0.5\int_0^t v(\tau)\,d\tau = -10$$

If this equation is differentiated and rearranged, the resulting second-order differential equation is

$$\frac{d^2v}{dt^2} + 0.05\frac{dv}{dt} + 0.125v(t) = 0$$

This is a homogeneous equation, so the complete solution will be the solution of this equation since there is no forcing function. To determine the form of the solution, the value of $b^2 = 0.0025$ while $4c = 0.5$. So $b^2 < 4c$ and the solution will be from case III and given by Eq. (3-31) with Eqs. (3-32) and (3-33). The damping coefficient is given from Eq. (3-32),

$$\alpha = \frac{b}{2} = 0.025$$

and the angular frequency is given by Eq. (3-33),

$$\omega = \frac{(4c - b^2)^{1/2}}{2} = \frac{(0.5 - 0.0025)^{1/2}}{2}$$

$$= 0.35$$

Then the solution can be written

$$v(t) = e^{-0.025t}[K_1 \sin (0.35t) + K_2 \cos (0.35t)]$$

The values of the two undetermined constants will be computed from the initial values of $v(t)$ and its derivative, $v'(0)$. In the initial condition the inductor carried a dc current of 2 A. The inductor does not have any dc voltage drop (since the voltage depends on the rate of change of the current), so there is no initial voltage across the capacitor, $v_C(0) = 0$, and this is also the node voltage. This gives one of the initial conditions, $v(0) = 0$. To get the other, note that $i_C(t) = C(dv/dt)$, so if the initial capacitor current were known, the voltage derivative would also be known. This initial condition can be found by an application of KCL just after the switch closes. Then, using the sign convention that currents directed into the node are positive and those directed out are negative, we have

$$-10 - i_R(0) + 2 - i_C(0) - i_L(0) = 0$$

but it has been shown that $i_R(0) = 0$ and $i_L(0) = 2$, so this equation becomes

$$i_C(0) = 4 \left.\frac{dv}{dt}\right]_{t=0} = -10$$

These two initial conditions are now applied to the solution above. Let us use the condition $v(0) = 0$ first,

$$v(0) = 0 = e^0[K_1 \sin (0) + K_2 \cos (0)] = K_2$$

So $K_2 = 0$. The other condition gives

$$4 \left.\frac{d}{dt} [K_1 e^{-0.025t} \sin (0.35t)]\right]_{t=0} = -10$$

or

$$4K_1[-0.025e^0 \sin (0) + 0.35e^0 \cos (0)] = -10$$

This gives us $1.4K_1 = -10$ or $K_1 = -7.14$. The final solution becomes

$$v(t) = -7.14e^{-0.025t} \sin (0.35t)$$ ■

These examples illustrate the basic procedures for finding the solution of the homogeneous part of the complete solution. Remember that in these examples,

TABLE 3–2

Forcing Function	Particular Solution
K	A
Kt	$A + Bt$
Ke^{-at}	Ae^{-at}
$\left. \begin{array}{l} K \sin(\omega t) \\ K \cos(\omega t) \end{array} \right\}$	$A \sin(\omega t) + B \cos(\omega t)$

there was no $H(t)$, so the complete solution is just the homogeneous part. In general, this will not be true.

Particular Solution. The particular solution will depend on the form of the forcing function, $H(t)$. There are a number of methods for finding this part of the solution, but the one best suited to network problems is called *substitution*. The idea here is that the particular solution will have the same *form* as the forcing function. So, knowing this, it is a matter of substituting such a solution into the differential equation and determining amplitudes to make the solution fit.

In electrical and electronic networks there are only a few forms that the forcing function can have, so it is quite reasonable to tabulate those forms for use in substitution. Table 3–2 lists the most common types of forcing functions and what particular solution should be used in each case for substitution. Notice that each has one or more unknown constants to be determined. These are *not* determined from initial conditions but rather from the differential equation itself. The following examples illustrate procedure.

EXAMPLE 3-14
The switch in the network of Fig. 3–26 closes at $t = 0$. Find the complete solution of the voltage.

SOLUTION
Since there is no power connected to the circuit in the $t = 0^-$ state, the initial conditions on the capacitor and inductor are $v_C(0) = 0$ and $i_L(0) = 0$. The

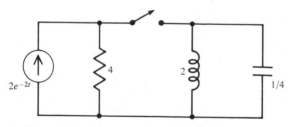

FIGURE 3–26. Network for Example 3-14.

differential equation is found from an application of KCL and the relations of Table 3–1.

$$\frac{d^2v}{dt^2} + \frac{dv}{dt} + 2v(t) = -16e^{-2t}$$

Clearly, there will be both homogeneous and particular parts to the solution of this problem. First the homogeneous solution will be found. The equation to be solved is

$$\frac{d^2v_h}{dt^2} + \frac{dv_h}{dt} + 2v_h(t) = 0$$

In this equation, $b^2 = 1$ and $4c = 8$, so the solution will be from case III with $\alpha = \frac{1}{2}$ and $\omega = (8 - 1)^{1/2}/2 = 1.3$. The solution is

$$v_h(t) = e^{-t/2}[K_1 \sin (1.3t) + K_2 \cos (1.3t)]$$

The values of the undetermined constants will be found by applying initial conditions to the complete solution, so the particular solution must be found first.

To find the particular solution, the form of the solution is obtained from Table 3–2. You can see that when the forcing function has the form of an exponential, the particular solution is given as $f_p(t) = Ae^{-at}$. So in this case the particular solution is $v_p(t) = Ae^{-2t}$. This is now substituted back into the differential equation and the derivatives indicated are computed. The result of this is

$$4Ae^{-2t} - 2Ae^{-2t} + 2A^{-2t} = -16e^{-2t}$$

This reduces to

$$4Ae^{-2t} = -16e^{-2t}$$

from which A must be equal to -4, so the particular solution is

$$v_p(t) = -4e^{-2t}$$

The complete solution is the sum of this particular solution and the homogeneous

$$v(t) = -4e^{-2t} + e^{-t/2}[K_1 \sin (1.3t) + K_2 \cos (1.3t)]$$

To find the values of the constants, the initial conditions of $v(0)$ and dv/dt at $t = 0$ will be used. Since there was no power connected in the initial state, it is clear that $v(0) = 0$. The other condition can be found from the observation that since $i_L(0) = 0$, all the current from the source must go through the capacitor when the switch closes, $i_C(0) = 2e^0 = 2$. Therefore,

$$\frac{1}{4}\frac{dv}{dt}\bigg]_{t=0} = 2 \quad \text{or} \quad \frac{dv}{dt}\bigg]_{t=0} = 8.$$

The first initial condition gives

$$v(0) = 0 = -4e^0 + e^0[K_1 \sin (0) + K_2 \cos (0)]$$

$$0 = -4 + K_2$$

Therefore,

$$K_2 = 4$$

To use the other condition it will be necessary to differentiate the complete solution:

$$8 = \frac{d}{dt}\left[-4e^{-2t} + e^{-t/2}[K_1 \sin (1.3t) + 4 \cos (1.3t)]\right]_{t=0}$$

$$= 8e^0 - \tfrac{1}{2}e^0[K_1 \sin (0) + 4 \cos (0)] + e^0[1.3K_1 \cos (0) - 1.3(4) \sin (0)]$$

$$= 8 - 2 + 1.3K_1$$

so that $K_1 = 1.54$. The complete solution can now be written as

$$v(t) = -4e^{-2t} + e^{-t/2}[1.54 \sin (1.3t) + 4 \cos (1.3t)]$$

A graph of this function appears in Fig. 3–27. ∎

EXAMPLE 3-15

The switch in the network of Fig. 3–28 closes at $t = 0$. Find an equation for the voltage across the 4-Ω resistor.

SOLUTION

Since there is but one current loop and the switch has been open, the initial conditions are $v_C(0) = 0$ and $i_L(0) = 0$. After the switch is closed, KVL can be used to obtain the following equation for the single current loop:

$$2 \sin (4t) - 8i(t) - 4\frac{di}{dt} - 2\int_0^t i(\tau)\,d\tau - 5 = 0$$

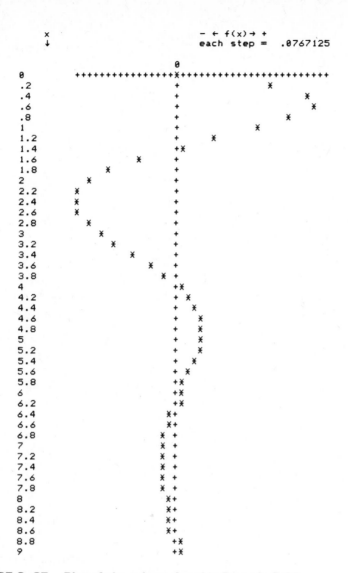

```
      x                              -  ← f(x) →  +
      ↓                             each  step =   .0767125

                                     θ
   θ           +++++++++++++++++++X+++++++++++++++++++++++++++
   .2                             +                    X
   .4                             +                         X
   .6                             +                         X
   .8                             +                    X
   1                              +                 X
   1.2                            +        X
   1.4                           +X
   1.6                 X         +
   1.8              X            +
   2           X                 +
   2.2       X                   +
   2.4       X                   +
   2.6       X                   +
   2.8          X                +
   3              X              +
   3.2              X            +
   3.4                 X         +
   3.6                    X      +
   3.8                       X  +
   4                           +X
   4.2                          +  X
   4.4                          +    X
   4.6                          +      X
   4.8                          +      X
   5                            +      X
   5.2                          +      X
   5.4                          +    X
   5.6                          + X
   5.8                         +X
   6                           +X
   6.2                         +X
   6.4                        X+
   6.6                        X+
   6.8                       X +
   7                         X +
   7.2                       X +
   7.4                       X +
   7.6                       X +
   7.8                       X +
   8                         X+
   8.2                       X+
   8.4                       X+
   8.6                       X+
   8.8                        +X
   9                          +X
```

FIGURE 3–27.   Plot of the voltage found in Example 3-14.

To obtain a differential equation, this equation is differentiated and rearranged. The result is

$$\frac{d^2i}{dt^2} + 2\frac{di}{dt} + \frac{1}{2}i(t) = 2\cos(4t)$$

This equation will have a homogeneous solution and a particular solution based on the form of the forcing function. First the solution of the homogeneous

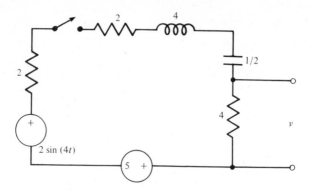

FIGURE 3–28. Network for Example 3-15.

equation will be found. The equation will be the same as that above but equal to zero on the right. Since $b = 2$, then $b^2 = 4$ and since $c = \frac{1}{2}$, then $4c = 2$. Thus $b^2 > 4c$, so the solution is from case I. From the definitions of case I the solution can be written

$$i_h(t) = K_1 e^{-1.7t} + K_2 e^{-0.3t}$$

For the particular solution, Table 3–2 shows that for the forcing function given above the appropriate particular solution is

$$i_p(t) = A \sin (4t) + B \cos (4t)$$

This is now substituted into the differential equation and the differentiation indicated is performed. The result is

$$(-15.5A - 8B) \sin (4t) + (8A - 15.5B) \cos (4t) = 2 \cos (4t)$$

This equation must be valid for all time, which can be possible only if the coefficients of the sine and cosine terms are equal independently. Since there is no sine on the right, the coefficient of sine on the left must be zero and the coefficient of cosine on the left must be equal to 2. This gives two equations for the amplitudes

$$-15.5A - 8B = 0$$

$$8A - 15.5B = 2$$

This simultaneous set of equations can easily be solved by Cramer's rule. The result is

$$A = 0.05 \quad \text{and} \quad B = -0.1$$

The complete solution is

$$i(t) = K_1 e^{-1.7t} + K_2 e^{-0.3t} + 0.05 \sin (4t) - 0.1 \cos (4t)$$

Now to find the two undetermined constants from the homogeneous solution this complete solution must be used. First it has been noted that $i(0) = 0$ at the beginning of the problem. Now for the derivative at $t = 0$ it will be necessary to consider the network at $t = 0^+$ together with the constraints of the capacitor and inductor conditions. There is no initial charge on the capacitor. Figure 3–29 shows the network drawn with all the conditions at $t = 0^+$. Since the current must still be zero (because of the inductor) there can be no voltage across the resistors. From the network you can see that there must be 5 V generated across the inductor the moment the switch is closed. So $v_L (0) = 4(di/dt) = -5$ V. The negative comes about because our defined current direction is clockwise, which would produce a voltage of the opposite polarity as that which is across the inductor. So the initial condition is $di/dt = -1.25$ at $t = 0$. Of course, this could also have been found by application of KVL to the network at $t = 0$. Then the KVL equation is

$$2 \sin (0) - 4i(0) - v_{4H} - v_C(0) - 4i(0) - 5 = 0$$

but, of course, $\sin (0) = 0$, $i(0) = 0$ and $v_C(0) = 0$. Then the KVL equation gives

$$v_{4H}(0) = 4i'(0) = -5 \text{ V}$$

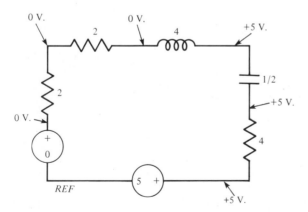

FIGURE 3–29. This is the network of Fig. 3–28 set up to find the initial conditions.

When the equation above is differentiated and $t = 0$ is substituted, the resulting expression is

$$\left(\frac{di}{dt}\right)_{t=0} = -1.25 = -1.7K_1 - 0.3K_2 + 0.2$$

The initial condition that $i(0) = 0$ gives

$$i(0) = 0 = K_1 + K_2 - 0.1$$

The following two equations result for the undetermined constants:

$$1.7K_1 + 0.3K_2 = 1.45$$

$$K_1 + K_2 = 0.1$$

These can be solved by simultaneous equation techniques to yield $K_1 = 1$ and $K_2 = -0.9$. The final solution is

$$i(t) = e^{-1.7t} - 0.9e^{-0.3t} + 0.05 \sin(4t) - 0.1 \cos(4t) \qquad \blacksquare$$

The previous examples have all involved simple circuits with only one loop or two nodes. The reason for this is that, in general, it is easier to obtain a solution of more complicated network problems using the method of Laplace transforms presented in Chapter 4. It will be instructive, however, to consider one example that includes controlled sources.

EXAMPLE 3-16
The network of Fig. 3–30 contains a controlled source between the capacitor and inductor. The switch closes at $t = 0$. Find an equation for the current after $t = 0$.

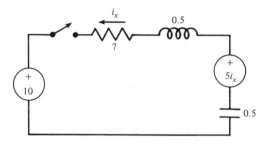

FIGURE 3–30.   Network for Example 3-16.

SOLUTION

First, it is clear that the capacitor is uncharged and that there is no current through the inductor in the initial state since the switch isolates the power source from these elements. KVL can be used to write the integrodifferential equation for the network

$$10 - 7i(t) - \frac{1}{2}\frac{di}{dt} - 5i_x(t) - 2\int_0^t i(\tau)\,d\tau = 0$$

It is clear that controlling current $i_x(t) = -i(t)$. When this is inserted and the equation is simplified, the result is

$$\frac{d^2i}{di^2} + 4\frac{di}{dt} + 4i(t) = 0$$

This is a homogeneous second-order differential equation. You can see that $b^2 = (4)^2 = 16$ and that $4c = 4(4) = 16$ also, so that the solution falls under case II. The solution can thus be written in the form of Eq. (3-30):

$$i(t) = K_1e^{-2t} + K_2te^{-2t}$$

The two constants are evaluated by finding $i(0)$ and $i'(0)$. Well, clearly $i(0) = 0$ since there can be initial current with the switch open. To find the derivative at $t = 0$, KVL can be applied just after the switch closes. Since $i(0) = 0$ the controlled source contributes nothing, nor does the resistor or the capacitor, which is uncharged. Thus

$$10 - \tfrac{1}{2}i'(0) = 0 \qquad \text{or} \qquad i'(0) = 20$$

Using these conditions gives the constants as follows. Using $i(0) = 0$;

$$K_1e^0 + K_2(0)e^0 = 0$$

Therefore

$$K_1 = 0$$

Now, using $i'(0) = 20$ gives

$$i'(t) = -2K_2te^{-2t} + K_2e^{-2t}$$

$$i'(0) = -2K_2(0)e^0 + K_2e^0 = 20$$

Therefore,

$$K_2 = 20$$

The complete solution is then

$$i(t) = 20te^{-2t}$$ ∎

This example illustrated both the occurrence of controlled sources into networks with capacitors and the occurrence of the special case of $b^2 = 4c$. The example shows that controlled sources present no difficulties. It is merely necessary to identify the dependency on network variables and then insert this into the equations. In this example $i_x = -i(t)$.

Another interesting feature of controlled sources can be seen if one assumes that the controlled source becomes $7i_x$ instead of $5i_x$ in Fig. 3–30. From Example 3-16 it is clear that the differential equation will become

$$\frac{d^2i}{di^2} + 4i(t) = 0$$

But in this equation $b = 0$. Therefore, the solution will be of case III with $\alpha = 0$ and $\omega = \sqrt{c}$. This means that there will be *no* damping of the oscillation. The controlled source has turned the network into an oscillator.

SUMMARY

In this chapter you have seen how the introduction of capactiors, inductors, and transformers gives rise to integrodifferential equations to solve a network. The following specific topics were identified:

1. When KVL is applied to a network that contains inductors, capacitors, or transformers, integrodifferential equations result for the currents.
2. When KCL is applied to a network that contains inductors, capacitors, or transformers, integrodifferential equations result for the node voltages.
3. Due to the energy storage property of capacitors and inductors, the initial conditions of these elements must be specified in a network problem. This requires that the time origin of the problem be established, often by a switch.
4. The initial condition of a capacitor is any charge, expressed as a voltage, present instantaneously before the time origin of the problem. Such a charge

is continuous and must therefore be the same just before and just after the time origin.

5. The initial condition of an inductor is any current flowing through the element instantaneously before the time origin of the problem. Such a current is continuous and must therefore be the same just before and just after the time origin.

6. The presence of inductors, capacitors, or transformers may give rise to linear first- or second-order differential equations with constant coefficients.

7. In each case the solution can be expressed as the sum of two parts, homogeneous and particular. The particular part depends on the form of the forcing function and the homogeneous part is the solution for no forcing function.

8. The solution of a first-order differential equation can be constructed to within one undetermined constant by an integration formula. The constant value can be found by matching the solution to one initial condition of the network.

9. The homogeneous solution of second-order differential equations can be determined to be one of three forms, depending on the relative values of the coefficients. There will be two undetermined constants.

10. The undetermined constants of the second-order differential equation solution are found by matching the complete solution to two initial conditions of the network.

PROBLEMS

3–1   The switch in the network of Fig. 3–31 closes at $t = 0$. Use KVL to find the integrodifferential equation for the current. Be sure to include initial conditions.

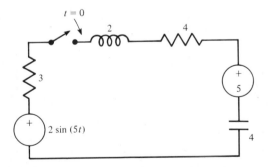

FIGURE 3–31. Network for Problems 3–1 and 3–17.

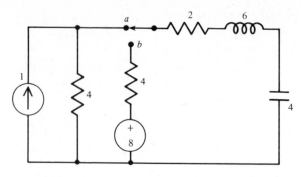

FIGURE 3–32.   Network for Problems 3–2 and 3–15.

3–2   Find the integrodifferential equations, using KVL, for the network of Fig. 3–32 when the switch has moved from position *a* to position *b* at $t = 0$.

3–3   Use KCL to find the integrodifferential equations for the network of Fig. 3–33.

3–4   The switch in the network of Fig. 3–34 closes at $t = 0$. Find the integrodifferential equation for the node voltage.

3–5   Find the initial conditions of the network in Fig. 3–35.

3–6   Find the set of integrodifferential equations for the network of Fig. 3–35 after the switch closes at $t = 0$.

3–7   Suppose that the switch in the network of Fig. 3–35 has been closed for a long time. The switch will now open again to create a new problem. What are the initial conditions?

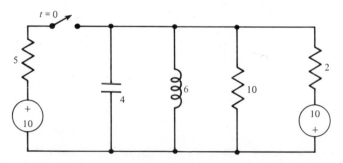

FIGURE 3–33.   Network for Problems 3–3 and 3–18.

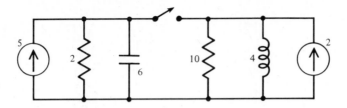

FIGURE 3–34. Network for Problems 3–4 and 3–16.

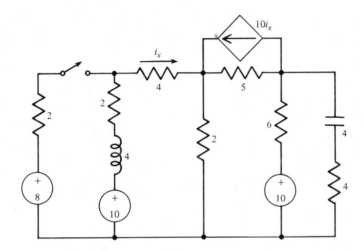

FIGURE 3–35. Network for Problems 3–5, 3–6, 3–7, and 3–8.

3–8 Find the set of equations that will solve the network of Fig. 3–35 under the conditions of Problem 3–7 if the switch opens at $t = 0$.

3–9 Use a transformer T model and KCL to set up the integrodifferential equation to solve the network of Fig. 3–36.

3–10 Use the direct transformer equations to set up the equations to solve the network of Fig. 3–36 using KVL.

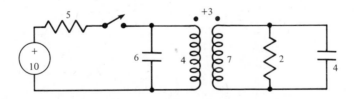

FIGURE 3–36. Network for Problems 3–9 and 3–10.

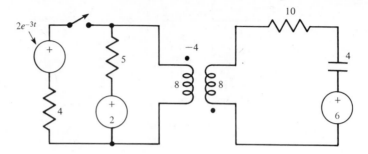

FIGURE 3–37.   Network for Problem 3–11.

3–11   Use KVL to find the equations that will solve the network of Fig. 3–37.

3–12   Consider the network of Fig. 3–38.
   (a) At $t = 0$ the switch moves from a to b. Find the complete solution for the current through the 4-F capacitor.
   (b) After 20 s the switch moves back to position a. Find the solution for the current through the 4-F capacitor again.
   (c) How much charge is left on the $\frac{1}{2}$-F capacitor?
   (d) Plot the voltage across the capacitor for $t = 0$ to $t = 60$ s.

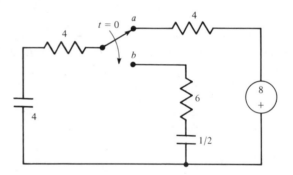

FIGURE 3–38.   Network for Problem 3–12.

3–13   At $t = 0$ the switch in the network of Fig. 3–39 closes. Find the solution for the voltage.

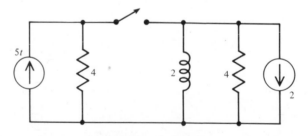

FIGURE 3–39.   Network for Problem 3–13.

CAPACITORS AND INDUCTORS

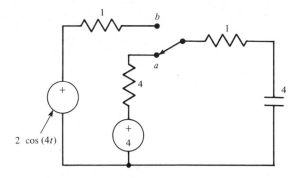

FIGURE 3–40. Network for Problem 3–14.

3–14 The switch in the network of Fig. 3–40 moves from *a* to *b* at *t* = 0. Find a solution for the voltage across the 4-F capacitor and plot to show the behavior.

3–15 Solve for the current in the network of Fig. 3–32 when the switch moves to position *b*.

3–16 Solve for the voltage of the network in Fig. 3–34 with the switch closed.

3–17 Calculate and plot the current in the network of Fig. 3–31.

3–18 Solve for the voltage in the network of Fig. 3–33.

3–19 Show that the following differential equations result for the loop currents of Fig. 3–41 when the switch closes at *t* = 0.

$$10i_1 + 5\frac{di_1}{dt} - 5i_2 = 10 \cos (3t)$$

$$-5i_1 + 5i_2 + 6\frac{di_2}{dt} = 0$$

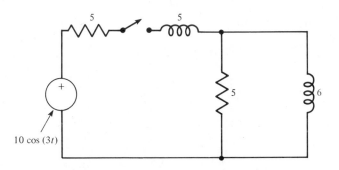

FIGURE 3–41. Network for Problem 3–19.

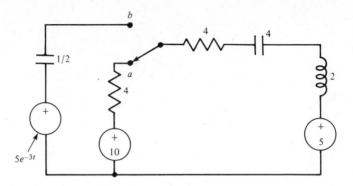

FIGURE 3–42. Network for Problem 3–20.

3–20 At $t = 0$ the switch in Fig. 3–42 moves from a to b. Find a complete solution for the current.

3–21 The switch in the network of Fig. 3–43 moves from a to b for 3 s and then back to a. Solve for the current through the inductor and plot for $t = 0$ to 10 s.

3–22 Solve the network of Fig. 3–30 if the controlled source is given by $7i_x$.

3–23 The set of equations given in Problem 3–19 can be solved by solving for i_1 from the second equation and substituting this into the first equation. The result will be a second-order differential equation for i_2. This can be solved and used to find i_1. Find the complete solutions for i_1 and i_2.

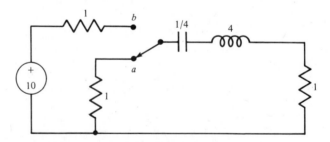

FIGURE 3–43. Network for Problem 3–21.

CHAPTER 4

Network Analysis with Laplace Transforms

Objectives

The objectives of this chapter can be stated in the form of a set of tasks or goals. At the conclusion of a study of this chapter and solving the problems at the end of the chapter, you will be able to:

1. Explain the concept of variable transformations and give a definition of the Laplace transformation.
2. Show how a table of time function and Laplace transform function pairs can be used to perform a transformation.
3. Show how a table of time operations and equivalent Laplace transform operations can be used to perform transformations.
4. Transform the integrodifferential equations resulting from network analysis to the equivalent Laplace equations.
5. Use partial fraction expansion to find the time function from a Laplace transform function.
6. Find the network variable time functions of any linear network using the methods of Laplace transforms.

INTRODUCTION

Laplace transformation represents a mathematical process by which any and all equations represented by the time variable t are converted into equations using a different variable, s. The motivation for doing this is that it turns out that problems to be solved in the t-representation are much easier to solve in the s-representation. In fact, all time functions, including transcendental, become algebra functions in s and *all* operations in time, including integration and differentiation, become *algebra* operations in s! Transformations such as this are very common in technical studies. The following section presents an analogy to another type of transformation with which you are already familiar so that you will understand the idea behind such transformations.

4-1.1 Concept of Transformations

In order to obtain a clear idea of the concept or idea behind variable transformations and to show how such transformation can be helpful in solving problems, it will be of value to consider another type, with which you are already familiar. This will be done by showing how a certain type of variable transformation can be used to simplify a problem.

The Problem. Suppose that a problem is given to measure the height of marks on a vertical pole, as shown in Fig. 4–1. Variable x measures these heights by a set of marks with the values $x_0, x_1, \ldots, x_5$. The answers required are the values of these marks in the units of x, such as meters.

Direct Solution. The direct way to solve this problem would be to use a distance-measuring instrument, such as a tape measure, to read off the distance from the base of the pole to each mark. Of course, if the pole is very high or

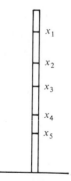

FIGURE 4–1. Problem to illustrate the concept of variable transformations.

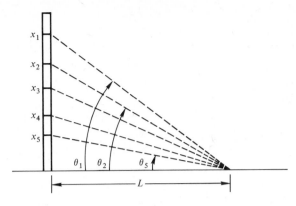

FIGURE 4–2. Pictorial display of the transformation to a variable set of angle to measure heights.

not strong enough to hold a climber, this could be very difficult. *Conceptually,* the problem is easy to solve, but *practically* there are some real difficulties.

Solution by Transformation. The problem will be much easier to solve if a transformation is made from the x variable to another variable, called angle θ. The transformation is made by using a *transformation law* to convert every value of x into a unique value of θ. The set of values of distance marks now becomes a set of values $\theta_1, \ldots, \theta_5$. Of course, you know now that we are talking about the trigonometric transformation. As shown in Fig. 4–2, it will be much easier to measure the angles each mark makes from some point a horizontal distance L from the base of the pole to the marks on the pole. Once the θ have been measured it will be necessary to perform the *inverse* transformation back to the x representation. This operation can be represented symbolically as

$$x = L \tan (\theta)$$

In actual practice, however, the measured values of θ are used in a table of θ and tan (θ) values. It is possible to use a mathematical expression to deduce tan (θ) from θ, but it is much more efficient to use the tables.

Summary. The process outlined above is an example of a variable transformation. The problem in x-space was transformed to a problem in θ-space and solved in that space. Then an inverse transformation was made of the solutions back to x-space. This same conceptual process is employed in Laplace transformations between the variable time and a Laplace variable.

4-1.2 Laplace Transformation

In Laplace transformations a problem in t-space is transformed into a problem in s-space, where s is the symbolic name of the Laplace variable. The problem is solved in s-space and then an inverse transformation is made of the solution back to t-space. The reason for making the transformation is that the problems are generally much easier to solve in s-space. For example, the integrodifferential equations are transformed directly to s-space, without conversion to a differential equation. Furthermore, the only initial conditions needed are the voltages across capacitors and the current through inductors, which are relatively easy to find. As the details of the procedure are studied, you should keep in mind this global objective of simplifying problem solution.

When a Laplace transformation of a network problem is performed, the variable t, all functions of t, $f(t)$, and all operations involving t, such as df/dt, are transformed into the new s-space.

The Laplace transformation can be defined symbolically as follows. For every time function $f(t)$, there is a unique function of the Laplace variable s, given by

$$F(s) = \mathcal{L}[f(t)] \tag{4-1}$$

where:

1. For every $f(t)$ there is one and only one $F(s)$. This means, at least formally, that if one of the two functions is known, the other is completely determined.
2. The transformation is always complete, so s and t can never appear in the same expression.

In the same fashion as Eq. (4-1), a formal definition of the inverse transformation can be stated as

$$f(t) = \mathcal{L}^{-1}[F(s)] \tag{4-2}$$

Both of these expressions are symbolic, just as tan (θ) is symbolic of a trigonometric transformation. It will turn out that the Laplace transformation, like the trigonometric, is easiest to use with a set of tables. However, there is value to seeing what the actual transformation law is and how the tables are developed from the law.

Laplace Transformation Law. The Laplace transform is one of a class of *integral transforms* and is defined by the equation

$$F(s) = \int_0^\infty f(t)e^{-st}\, dt \tag{4-3}$$

where

$$F(s) = \text{Laplace transform of } f(t)$$

$$f(t) = \text{time function}$$

$$s = \text{Laplace variable}$$

$$t = \text{time variable}$$

A common notation, and the one followed in this text, represents time functions by lowercase characters and the Laplace transform by the upper case characters. Notice that the Laplace transform is defined only for $t > 0$. This means that the concept of Laplace transform includes the idea of time origin which was introduced for problems with capacitors and inductors. It is important for you to remember that problems solved with Laplace transforms will provide valid results only for $t > 0$.

The following example illustrates how the definition of Eq. (4-3) is used to develop transform pairs in the construction of a table of pairs.

EXAMPLE 4-1

Find the Laplace transform of the time function given by $f(t) = 3e^{-2t}$ using the integral definition of Eq. (4-3).

SOLUTION
From Eq. (4-3),

$$F(s) = 3 \int_0^\infty e^{-2t} e^{-st} \, dt$$

$$F(s) = 3 \int_0^\infty e^{-(s+2)t} \, dt$$

But, of course, $\int e^{ax} \, dx = e^{ax}/a$, so that the integral becomes

$$F(s) = -\frac{3}{s+2} e^{-(s+2)t} \bigg]_0^\infty = -\frac{3}{s+2} (e^{-\infty} - e^0)$$

or

$$F(s) = \frac{3}{s+2} \qquad \blacksquare$$

You can see already in this example that the Laplace transform of a transcendental function of time resulted in an algebraic function. A result like that found in this example can be generalized to any exponential function of the same form,

$$\mathcal{L}[Ke^{-at}] = \frac{K}{s+a} \qquad (a > 0) \qquad (4\text{-}4)$$

The restriction that $a > 0$ is necessary because the integration performed in Example 4-1 will be valid only for a negative exponential. By performing integrations such as that of Example 4-1 on the typical functions found in network analysis, it is possible to construct a table of transform pairs such as the one given by Eq. (4-4).

Inverse Transformation.   The inverse transformation of a Laplace transform to find the time function is also defined by an integration process but of a more complicated nature. You will see later that the Laplace variable can be a complex number, $s = \sigma + j\omega$, so that integration over this variable requires integration in the complex plane. Instead of showing such an integration, even as an example, it is more constructive to note that the transformations found from Eq. (4-3) are reversible. For example, from Eq. (4-4) it is possible to define the inverse transformation of

$$\mathcal{L}^{-1}\left[\frac{K}{s + a}\right] = Ke^{-at} \tag{4-5}$$

This means that whenever the Laplace transform has the form given on the left side of Eq. (4-5), the time function will be an exponential of the form given.

EXAMPLE 4-2

Find the time function of the following Laplace transform,

$$F(s) = \frac{3.79}{s + 4.4}$$

SOLUTION

Since this transform is clearly of the form of Eq. (4-5), the solution can be easily written, with $a = 4.4$ and $K = 3.79$,

$$f(t) = 3.79e^{-4.4t} \qquad \blacksquare$$

Laplace Transforms of Operations.   One of the most attractive features of using Laplace transform to solve network problems is that math operations in time generally transform into simple multiplication and division operations in s-space. Just as there are tables of time functions and Laplace functions, there are tables of t-space operations and s-space operations. The following example illustrates how such operation pairs are established.

EXAMPLE 4-3

Find the Laplace transform operation equivalent to differentiation in time,

$$g(t) = \frac{df}{dt}$$

SOLUTION

To solve this problem, let us use the basic definition of a Laplace transform as given by Eq. (4-3) and write down the transform of $g(t)$:

$$G(s) = \int_0^\infty g(t)e^{-st}\,dt = \int_0^\infty \frac{df}{dt}\,e^{-st}\,dt$$

The last integral can be simplified using integration by parts, $\int u\,dv = uv - \int v\,du$. For the integral above, we take $dv = (df/dt)\,dt$, which means that $f = v$ and $u = e^{-st}$, so that $du = -se^{-st}$. The expression for $G(s)$ now becomes

$$G(s) = f(t)e^{-st}\Big]_0^\infty + \int_0^\infty se^{-st}f(t)\,dt$$

or

$$G(s) = -f(0) + s\int_0^\infty f(t)e^{-st}\,dt$$

But you can recognize the integral in this last expression as just the Laplace transform of the time function $f(t)$. So the general result is

$$\mathcal{L}\left[\frac{df}{dt}\right] = sF(s) - f(0) \qquad\blacksquare$$

This example shows that the operation of differentiation in time becomes multiplication by s in s-space. The $f(0)$ term shows that the *initial conditions* will be built into the transformation process. From evaluations such as this for other operations, it is possible to construct a table of operation transformations.

4-2

LAPLACE TRANSFORM TABLES

All the functions and operations typically found in network analysis have been known and studied for many years. Consequently, the Laplace transforms of these have been developed and tabulated many times. When the objective is to use Laplace transforms as a tool for network analysis, there is really not much point in redoing the integrals. Instead, transform tables will be used which give all the standard network analysis functions and all the standard network analysis time operations. In this section the use of the tables will be presented.

4-2.1 Function Tables

A tabulation of all the common functions of time found in network analysis problems and their corresponding Laplace transforms are presented in Table 4–1. The entries in this table can be used to find the Laplace transform from a given time function or a time function from a given Laplace transform.

Finding the Transform. The table is arranged in two columns, one with the time function and the other with the corresponding Laplace transforms. Each time function has one or more constants associated with it and the Laplace transforms show how these constants appear in the transform. All of the transform pairs are valid for $t > 0$ only. This is equivalent to saying that each time function in the table can be considered to be multiplied by the step function, $u(t)$. The following example illustrates how transforms are found from time functions using the table.

TABLE 4–1 Laplace Transforms of Functions

	$f(t)$	$F(s)$
1.	$K\delta(t)$	K
2.	K	$\dfrac{K}{s}$
3.	Kt	$\dfrac{K}{s^2}$
4.	Ke^{-at} $(a > 0)$	$\dfrac{K}{s + a}$
5.	Kte^{-at} $(a > 0)$	$\dfrac{K}{(s + a)^2}$
6.	$K \sin (\omega t)$	$\dfrac{K\omega}{s^2 + \omega^2}$
7.	$K \cos (\omega t)$	$\dfrac{Ks}{s^2 + \omega^2}$
8.	$Ke^{-at} \sin (\omega t)$ $(a > 0)$	$\dfrac{K\omega}{(s + a)^2 + \omega^2}$
9.	$Ke^{-at} \cos (\omega t)$ $(a > 0)$	$\dfrac{K(s + a)}{(s + a)^2 + \omega^2}$

EXAMPLE 4-4

Find the Laplace transforms of the following time functions.

$$\text{(a) } 29.7 \quad \text{(b) } -16t \quad \text{(c) } 4 \cos{(2t)} \quad \text{(d) } 6e^{-3t} \sin{(8t)}$$

SOLUTION

(a) From Table 4–1, entry 2,

$$\mathcal{L}[29.7] = \frac{29.7}{s}$$

(b) From Table 4–1, entry 3,

$$\mathcal{L}[-16t] = -\frac{16}{s^2}$$

(c) From Table 4–1, entry 7,

$$\mathcal{L}[4 \cos{(2t)}] = \frac{4s}{s^2 + 4}$$

(d) From Table 4–1, entry 8

$$\mathcal{L}[6e^{-3t} \sin{(8t)}] = \frac{48}{(s + 3)^2 + 64}$$

Generally speaking, finding the transform of a time function is a simple matter of looking up the form of the time function in the table and writing down the transform with appropriate constants.

Finding the Time Function. The process of finding the time function from a Laplace transform is often called *inversion* because it is the inverse operation of finding the transform. Generally, it is more difficult, as you will see later, because the form of the transform will often not be exactly like that of the table. The following examples illustrate cases where the inversion is obvious from Table 4–1.

EXAMPLE 4-5

Find the time function of each of the following Laplace transforms.

$$\text{(a) } \frac{6}{2s} \quad \text{(b) } \frac{5}{s + 3} \quad \text{(c) } \frac{12}{2s + 4} \quad \text{(d) } \frac{6s + 12}{(s + 2)^2 + 16}$$

The function pairs of Table 4–1 can be used to find the time functions of all of these.

SOLUTION

(a) $\mathcal{L}^{-1}\left[\dfrac{6}{2s}\right] = 3$

(b) $\mathcal{L}^{-1}\left[\dfrac{5}{s + 3}\right] = 5e^{-3t}$

(c) This transform is not in the table in this form.
However, if the denominator is factored, $2s + 4 = 2(s + 4)$, then the transform is of entry 4,

$$\mathcal{L}^{-1}\left[\frac{12}{2(s + 2)}\right] = \mathcal{L}^{-1}\left[\frac{6}{s + 2}\right] = 6e^{-2t}$$

(d) This transform has a denominator like some of those in the table but the numerator is wrong. Again, by factoring, $6s + 12 = 6(s + 2)$ and now the transform looks like entry 9,

$$\mathcal{L}^{-1}\left[\frac{6(s + 2)}{(s + 2)^2 + 16}\right] = 6e^{-2t}\cos(4t)$$ ∎

As this example illustrates, the transforms can often be expressed in a way that disguises their presence in the table. One of the difficulties involved in using Laplace transforms to solve network problems is getting the transform into a form by which its time function can be recognized from the table. This is particularly true for the quadratic forms.

Quadratic Forms. Table 4–1 shows that quadratic denominators demonstrate the presence of sine or cosine terms in the time functions. This turns out to be true *only* for quadratic denominators which have *complex roots*. To see this, consider the last transform of Example 4-5. The denominator could have been written in the form $(s + 2)^2 + 16 = s^2 + 4s + 20$. In this form it could not have been recognized from the table. If we used the quadratic formula to factor the quadratic, the result would be

$$r = \frac{-4 \pm [4^2 - 4(20)]^{1/2}}{2} = -2 \pm 4j$$

This shows that the two roots, $-2 + 4j$ and $-2 - 4j$, are complex. Then the quadratic equation above can be written

$$s^2 + 4s + 20 = (s + 2 - 4j)(s + 2 + 4j)$$

or, if partially multiplied out,

$$s^2 + 4s + 20 = (s + 2)^2 + 16$$

So, you see, we have come full circle and shown that a quadratic with complex roots can be written in the form shown in Table 4–1 for sine and cosine functions. Notice that the real part of the complex root is the exponential constant a, and that the complex part of the root is the angular frequency ω. *However*, all quadratics do not have complex roots, and will therefore not represent sine and cosine functions. Consider the following transform:

$$F(s) = \frac{24}{s^2 + 9s + 18}$$

Let us factor the denominator, as above:

$$r = \frac{-9 \pm [9^2 - 4(18)]^{1/2}}{2} = -4.5 \pm 1.5$$

The roots are -3 and -6, so $s^2 + 9s + 18 = (s + 3)(s + 6)$. The roots are *real* and the denominator cannot be put in the form of the sine and cosine terms of the table. In fact, the transform now has the form

$$F(s) = \frac{24}{(s + 3)(s + 6)}$$

This is not even in the table. The next section will show how transforms of this type are inverted to find the time function.

In summary, any quadratic denominator must be factored. If the roots are complex, the denominator can be put in the form of one of the sine or cosine time functions of the table. If the roots are real, other evaluational techniques to be presented later in this chapter will be used.

4-2.2 Operation Table

In Table 4–2 the typical t-space operations encountered in network analysis are presented together with the corresponding operations in s-space. This table will be used to reduce network equations developed from KVL or KCL to forms that provide easy solution by algebraic means. The table is arranged in columns, with O_t representing operations in time performed on some function $f(t)$, and O_s to represent the corresponding operations in s-space on the transform of the time function $F(s)$.

TABLE 4–2 Laplace Transforms of Operations

	$O_t[f(t)]$	$O_s[F(s)]$
1.	$f_1(t) + f_2(t)$	$F_1(s) + F_2(s)$
2.	$Kf(t)$	$KF(s)$
3.	$\dfrac{df}{dt}$	$sF(s) - f(0)$
4.	$\dfrac{d^2f}{dt^2}$	$s^2F(s) - sf(0) - \dfrac{df}{dt}\bigg]_{t=0}$
5.	$\displaystyle\int_0^t f(\tau)\,d\tau$	$\dfrac{F(s)}{s}$

Entry 1: Association. The first entry in the table expresses the important fact that if the Laplace transform of a sum or difference of time functions is taken, the result is the sum or difference of the individual Laplace transforms. This means that the Laplace transform of an entire equation can be taken by substituting the Laplace transform of each individual term in the equation.

Entry 2: Constant Product. If a time function is multiplied by some constant, the Laplace transform of this product will be the same constant multiplied times the transform of the time function. In other words, constant multipliers factor through the transform operation.

Entry 3: First Derivative. As noted in Example 4-3, the Laplace transform of the derivative of a function of time is the transform of the time function multiplied by s. In addition, the initial value of the time function is subtracted from the transform. This shows that differentiation becomes multiplication by s and that the initial conditions are built into the Laplace transform operation.

Entry 4: Second Derivative. If the Laplace transform of a second derivative is taken, the result is seen to be s^2 times the transform of the time function. In this case the initial conditions consist of the time function and its time derivative evaluated at $t = 0$.

The Laplace transform of higher-order derivatives can be expressed by successively higher order products by s. In network analysis there is seldom need for more than second-order derivative transformations, so that is the highest order presented in Table 4–2.

Entry 5: Integration. This term shows that if the Laplace transform is taken of the integral of a time function, the result is simply the transform of the time function divided by s. In a sense the operation of integration becomes division by s. It is important to remember that any initial condition on the integral will have already been identified as the value of the integral at $t = 0$ and expressed as a separate, constant term. The following example illustrates this concept.

EXAMPLE 4-6

The switch in the network of Fig. 4–3 moves from position a to position b at $t = 0$. Find the current through the resistor.

SOLUTION

Let us use KVL to set up an equation for the current for $t > 0$. Since there is only the capacitor and resistor, the voltage sum is

$$v_{2F} + v_{4\Omega} = 0$$

Using the results of Chapter 3, this can be written

$$v_{2F}(0) + \tfrac{1}{2} \int_0^t i(\tau)\, d\tau + 4i(t) = 0$$

Now, a study of the circuit for $t < 0$ shows that the capacitor will have charged up to 16 V in position a. Also, given the clockwise direction of the assigned current, this 16 V will have the opposite polarity of the voltage drops due to the current. The final equation is

$$-16 + \tfrac{1}{2} \int_0^t i(\tau)\, d\tau + 4i(t) = 0$$

By the previous methods of differential equations, we would differentiate this equation. Instead, let us use the Laplace transforms of Tables 4–1 and 4–2 to convert this equation to s-space. According to entry 1 of Table 4–2, the trans-

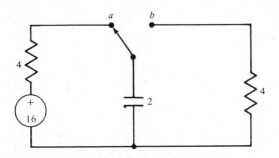

FIGURE 4–3. Network for Example 4-6.

form of the equation is the transform of the individual terms. Notice that the initial condition of the integral term has already been identified (the -16) and expressed as a separate term. The following operations show the development of a transform of the current from the equation above.

$$\mathscr{L}[-16] + \mathscr{L}\left[\tfrac{1}{2}\int_0^t i(\tau)\,d\tau\right] + \mathscr{L}[4i(t)] = 0$$

$$-\frac{16}{s} + \frac{I(s)}{2s} + 4I(s) = 0$$

The first term came from entry 2 of Table 4–1, the second from entry 5 of Table 4–2, and the third from entry 2 of Table 4–2. Solving for $I(s)$,

$$\left(4 + \frac{0.5}{s}\right)I(s) = \frac{16}{s} \quad \text{or} \quad I(s) = \frac{16}{s(4 + 0.5/s)}$$

Simplifying the last expression gives

$$I(s) = \frac{4}{s + 0.125}$$

This expression for the Laplace transform of the current can be found directly from Table 4–1. Thus the current solution is

$$i(t) = 4e^{-0.125t} \qquad \blacksquare$$

This example illustrates the general procedure to be followed to solve a network analysis problem with Laplace transforms. These steps can be summarized as follows:

1. Determine the initial conditions on capacitors, inductors, and transformers.
2. Use KVL or KCL, along with the i–v relationships of Table 3–1 to find an equation or set of integrodifferential equations for the network variables.
3. Transform the equation or equations into s-space using Tables 4–1 and 4–2.
4. Use standard algebra techniques, including Cramer's rule and determinants if necessary, to solve for the unknown or unknowns desired, as their Laplace transforms.
5. Invert the Laplace transforms to find the corresponding time functions.

EXAMPLE 4-7

In the network of Fig. 4–4 the switch moves from a to b at $t = 0$. The switching from position a to b is instantaneous, so the initial conditions on the inductor are not compromised. Find the current for $t > 0$.

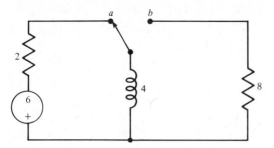

FIGURE 4–4. Network for Example 4-7.

SOLUTION

Let us just follow the steps outlined above exactly as stated.

1. The initial conditions are established with the switch in position a. Clearly, there is a dc current through the inductor given by Ohm's law as, $i_{4H}(0) = 6/2 = 3$ A, directed up through the inductor.

2. Using KVL, the equation is easily found since in position b there are only the inductor and the resistor,

$$4 \frac{di}{dt} + 8i(t) = 0$$

3. Now the Laplace transforms of each term above is found from the tables, giving

$$4sI(s) - 4i(0) + 8I(s) = 0$$

Notice that the transform of the derivative includes the initial condition on the current. This has been found in step 1, so it is inserted into the equation. After rearrangement the equation becomes

$$(4s + 8)I(s) = 12$$

$$I(s) = \frac{12}{4s + 8} = \frac{3}{s + 2}$$

This Laplace transform is in the tables directly, that the solution can be written as

$$i(t) = 3e^{-2t} \qquad \blacksquare$$

Products.   It is very important for you to realize that products of time functions *do not* transform into products of Laplace transforms, and vice versa.

Thus, if a time function has the form $f(t) = 5te^{-6t}$, you *cannot* find the transform of $5t$ and the transform of e^{-6t} and multiply them together. Coming the other way, if you have a transform given by

$$\frac{2s}{s^2 + 9} \frac{5}{s + 8} = \frac{10s}{(s^2 + 9)(s + 8)}$$

you *cannot* find the time functions of the two transforms on the left and consider the product to be the complete time function.

To find the Laplace transform of a product of time functions, it is necessary to use a table which shows the form explicitly or to integrate the function directly. In network analysis the form of the time functions encountered is covered by the entries in Table 4–1.

To find the time function of a product of transforms is a little more complicated. The reason for this is that the transform that results from solving a set of simultaneous equations in s-space is often the sum of many transforms, placed over a common denominator. In this form they may appear to be a product of transforms. To find the time function it will be necessary to break the transform up into sums and differences of its individual terms, each of which can then be inverted using Table 4–1. The procedure for doing this is called partial fraction expansion and is described in the next section.

4-3

INVERTING LAPLACE TRANSFORMS: PARTIAL FRACTION EXPANSION

The process of finding the Laplace transform of network time functions and network equations is quite easy to carry out using Tables 4–1 and 4–2 in conjunction with the steps outlined previously. The last step, inverting the transform to find the time function, is usually not so easy. The reason for this is that the transforms which result from analysis of a network are often a *composite* of several simple functions.

To see how this is a problem, suppose that the result of finding a current in some network gave a transform of

$$I(s) = \frac{4}{s + 2} + \frac{3}{s + 6} \tag{4-6}$$

Using Table 4–1, it is easy to see that the time function is given by

$$i(t) = 4e^{-2t} + 3e^{-6t} \tag{4-7}$$

Now, suppose that another problem gave a transform of

$$I(s) = \frac{7s + 30}{s^2 + 8s + 12} \qquad (4\text{-}8)$$

But this is *exactly* the same transform as Eq. (4-6) and has exactly the same time function as given by Eq. (4-7)! To see this, it is only necessary to put the transform of Eq. (4-6) over a common denominator,

$$I(s) = \frac{4}{s + 2}\frac{s + 6}{s + 6} + \frac{3}{s + 6}\frac{s + 2}{s + 2}$$

$$= \frac{4(s + 6) + 3(s + 2)}{(s + 2)(s + 6)}$$

$$= \frac{7s + 30}{s^2 + 8s + 12}$$

From this you can see that the transforms found as a result of network analysis will often be in a mixed form. It will be necessary to break them up into sums and differences of simpler terms before the table can be used to find the time functions. There are a number of ways in which this can be accomplished. This section presents a method called *partial fraction expansion*.

The following example illustrates a typical problem and the form that a solved transform may take in the solution.

EXAMPLE 4-8
Find the Laplace transform for the voltage in the network of Fig. 4–5. The switch closes at $t = 0$.

SOLUTION
The first step is to find the initial conditions on the inductor and capacitor. In the initial state all of the 2-A source must go through the inductor since it is a

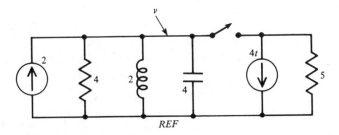

FIGURE 4–5. Network for Example 4-8.

dc short circuit. Thus $i_{2H}(0) = 2$ A, directed down. There is no voltage drop across the inductor in the dc state, so $v_{4F}(0) = 0$. The next step is to find the equations. Use of KCL seems appropriate since the elements are in parallel. The equation that results is

$$0.45v(t) + \frac{1}{2}\int_0^t v(\tau) \, d\tau + 4\frac{dv}{dt} = -4t$$

Taking Laplace transforms of the equation gives us

$$0.45V(s) + \frac{1}{2s} V(s) + 4sV(s) - 4v(0) = -\frac{4}{s^2}$$

Now, using $v(0) = 0$ and grouping gives

$$\left(0.45 + \frac{1}{2s} + 4s\right)V(s) = -\frac{4}{s^2}$$

Solving for $V(s)$ gives the final result,

$$V(s) = -\frac{4}{s^2(0.45 + 0.5s^{-1} + 4s)} \qquad \blacksquare$$

It is clear that no transform like that which resulted from Example 4-8 is given in Table 4–1.

4-3.1 Preliminary Preparation
Before the technique of partial fraction expansion can be applied to a Laplace transform, certain preliminary operations on the transform are necessary. These are required to take the "raw" transform which comes from solution of the network equations, like that of Example 4-8, and place it in a form suitable for partial fraction expansion.

Polynomial Ratio. First the transform must be expressed as the ratio of two polynomials in positive powers of s. All transforms encountered in linear electrical and electronic networks can be expressed as the ratio of two such polynomials. This can be defined by the expression

$$G(s) = \frac{C(s)}{D(s)} \qquad (4\text{-}9)$$

where

$$G(s) = \text{Laplace transform}$$

$$C(s) = c_r s^r + c_{r-1} s^{r-1} + \cdots + c_2 s^2 + c_1 s + c_0 \qquad (4\text{-}10)$$

$$D(s) = s^n + d_{n-1} s^{n-1} + \cdots + d_2 s^2 + d_1 s + d_0 \qquad (4\text{-}11)$$

Equation (4-9) shows the transform expressed as a ratio of the two polynomials given by Eq. (4-10) for the numerator and Eq. (4-11) for the denominator.

Notice that the coefficient of the highest power of s in the denominator s^n has been expressed as unity, $d_n = 1$. This is done to make later calculations a little easier and to avoid mistakes. If the coefficient is not 1, as often happens, it can be factored from the denominator and divided into each coefficient of the numerator, as shown in the next example.

In general, the coefficients in the numerator will be positive or negative real numbers or zero. The coefficients in the denominator may, in a general mathematical sense, be positive or negative real numbers or zero, *but* in linear network analysis they will never be negative. This fact is often helpful in finding errors in computation, since if a negative is present an error has been made.

Transforms will almost never have the explicit form given by Eq. (4-9), as derived from the equations. It will be necessary to perform algebraic manipulations to place the transform in this representation. This is called placing the transform in *good form*.

EXAMPLE 4-9

Put the transform found in Example 4-8 in good form.

SOLUTION

This transform is not already in good form because the denominator contains a term in s^{-1}. This can be eliminated by multiplying the term s^2 into the denominator:

$$V(s) = -\frac{4}{0.45s^2 + 0.5s + 4s^3}$$

Now a 4 is factored from the s^3 term in the denominator so that the highest power has a unity coefficient:

$$V(s) = -\frac{1}{s^3 + 0.1125s^2 + 0.125s} \qquad \blacksquare$$

The transform is now in good form as the ratio of two polynomials. In the numerator there is only $c_0 = 1$, while the denominator has $d_3 = 1$, $d_2 = 0.1125$, $d_1 = 0.125$, and $d_0 = 0$.

EXAMPLE 4-10

Express the following Laplace transform in good form:

$$I(s) = \frac{6s(s - 3)}{(2 + 4s^{-1})(s + 3)(s + 4)}$$

SOLUTION

The objective can be met by simply multiplying out numerator and denominator until only polynomial terms remain. This gives

$$I(s) = \frac{6s^2 - 18s}{2s^2 + 18s + 52 + 48s^{-1}}$$

To eliminate the s^{-1} term we can multiply top and bottom by s:

$$I(s) = \frac{6s^3 - 18s^2}{18s^2 + 52s + 2s^3 + 48}$$

Now a 2 is factored from the s^3 term and the result is rearranged to yield

$$I(s) = \frac{3s^3 - 9s^2}{s^3 + 9s^2 + 26s + 24} \qquad ∎$$

Highest Power Test. Partial fraction expansion can be used only on a transform that satisfies a certain test of the powers of the polynomials in numerator and denominator. The test can be stated in the form of a rule:

Power Test Rule. Partial fraction expansion can be performed only on a polynomial ratio for which the highest power of s in the numerator is less than the highest power of s in the denominator.

With reference to the polynomial expression in Eq. (4-9), this would be stated as

$$r < n$$

In most cases the transforms that result from network analysis will satisfy this test. In those instances when the test is not satisfied, it will be necessary to divide the fraction until the remainder does satisfy the highest power test. Partial fraction expansion is then carried out on the remainder.

In a most general sense, if we divide $D(s)$ into $C(s)$ the result can be expressed as

$$G(s) = g_0 + g_1 s + \cdots + g_j s^j + F(s) \qquad (4\text{-}12)$$

where

$$F(s) = \frac{A(s)}{D(s)} \qquad (4\text{-}13)$$

and

$$A(s) = a_m s^m + a_{m-1} s^{m-1} + \cdots + a_1 s + a_0 \qquad (4\text{-}14)$$

The remainder, $F(s)$, is the ratio of two polynomials which does satisfy the highest power test, so $m < n$. The constants g_0 to g_j are positive or negative real numbers or zero. They can be inverted and become impulse or *delta functions*, which were discussed in Chapter 1. These terms appear in network analysis to represent initial condition impulses occurring with closing and opening of switches.

EXAMPLE 4-11

Show that the transform of Example 4-10 does not satisfy the highest power test. Perform necessary divisions to express the transform in the form of Eq. (4-12).

SOLUTION

The result of Example 4-10 had a numerator with a term s^3 and a denominator with a term s^3. Since the highest powers are the *same*, the test $r < n$ is not satisfied. Dividing gives us

$$
\begin{array}{r}
3 \\
s^3 + 9s^2 + 26s + 24 \overline{\smash{\big)}\ 3s^3 - 9s^2 } \\
\underline{-3s^3 - 27s^2 - 78s - 72} \\
-36s^2 - 78s - 72
\end{array}
$$

Now the remainder satisfies the power test since it is only in s^2. The result is

$$I(s) = 3 - \frac{36s^2 + 78s + 72}{s^3 + 9s^2 + 26s + 24} \qquad \blacksquare$$

When finding the time function of the transform of this example you can see that the first term will be $3\,\delta(t)$, since the inverse of a constant is an impulse or delta function.

Polynomial Roots. Recall the following properties of polynomials:

1. A polynomial $P(x)$ or order k will have k roots defined by finding k values of x for which $P(x) = 0$. The roots may be real or complex numbers.
2. A polynomial can always be written as a product of factors formed by the difference between the variable and the roots. If r_1, r_2, . . ., r_k are the k roots of the polynomial $P(x)$, then

$$P(x) = a_k(x - r_1)(x - r_2) \cdots (x - r_k) \qquad (4\text{-}15)$$

where a_k is the coefficient of x^k in the polynomial.

Now, the polynomial in the numerator of $F(s)$ as defined by Eq. (4-14) is of order m and therefore has m roots. Let z_1, z_2, . . ., z_m be the m roots. Then $A(s)$ can be written as

$$A(s) = a_m(s - z_1)(s - z_2) \cdots (s - z_m) \qquad (4\text{-}16)$$

Also, the denominator is a polynomial, $D(s)$, of order n. Let p_1, p_2, . . ., p_n represent the n roots of this polynomial. Then it can be written in the form

$$D(s) = (s - p_1)(s - p_2) \cdots (s - p_n) \qquad (4\text{-}17)$$

Note that there is no coefficient in front of the factor product in Eq. (4-17), since the denominator was adjusted such that that coefficient was unity.

Using Eqs. (4-16) and (4-17), it is now possible to write the transform defined in Eq. (4-13) as

$$F(s) = \frac{a_m(s - z_1)(s - z_2) \cdots (s - z_m)}{(s - p_1)(s - p_2) \cdots (s - p_n)} \qquad (4\text{-}18)$$

This is the factored form of the Laplace transform. The roots of the numerator, z_1, z_2, . . ., z_m, are called *zeros* of the transform because they represent values of s that will cause the transform to equal zero:

$$F(z_i) = 0$$

where z_i is any of the m roots of $A(s)$.

The roots of the denominator, p_1, p_2, . . ., p_n, are called the *poles* of the transform. If s ever takes on one of the values of the poles, there will be a factor equal to zero in the denominator, so the transform will go to infinity:

$$\lim_{s \to p_i} [F(s)] = \infty$$

where p_i is any of the n roots of $D(s)$.

The last step in the preparation, then, is to find the poles (roots) of the denominator and express the denominator in terms of the product of the factors. Although there is physical significance to the zeros, it is not necessary to write the numerator in terms of the zeros to perform partial fraction expansion. In fact, it makes the process somewhat more difficult to do so. Thus the last step of the preparation leaves the transform in the partially factored form,

$$F(s) = \frac{a_m s^m + a_{m-1} s^{m-1} + \cdots + a_1 s + a_0}{(s - p_1)(s - p_2) \cdots (s - p_n)} \tag{4-19}$$

The transform is now ready for partial fraction expansion.

The following examples illustrate the process of expressing the Laplace transform in the form of Eq. (4-18) or (4-19).

EXAMPLE 4-12

Express the transform of Eq. (4-8) as factored products. Identify the poles and zeros.

SOLUTION

For the numerator of Eq. (4-8) the polynomial is

$$A(s) = 7s + 30 = 7(s + 30/7)$$

Therefore, there is one zero at $z_1 = -30/7$ since this is the value of s that will make the polynomial equal to zero. The denominator of Eq. (4-8) has already been factored and can be written

$$D(s) = s^2 + 8s + 12 = (s + 2)(s + 6)$$

Thus it is clear there are two poles, $p_1 = -2$ and $p_2 = -6$, because there are values of s that will make the polynomial equal to zero. In full factored form the transform is

$$I(s) = \frac{7(s + 30/7)}{(s + 2)(s + 6)}$$ ∎

EXAMPLE 4-13

Express the transform of Example 4-9 in factored form. Identify the poles and zeros.

SOLUTION

From the transform found in Example 4-9 it is clear that there are no zeros since $A(s) = -1$. The poles are found by finding the roots of the cubic in the denominator:

$$D(s) = s^3 + 0.1125s^2 + 0.125s = 0$$

First an s can be factored from each term,

$$s(s^2 + 0.1125s + 0.125) = 0$$

This shows that one root is for $s = 0$; therefore, $p_1 = 0$. The remaining two roots are found from the quadratic formula,

$$\text{poles} = \text{roots} = \frac{-0.1125 \pm [0.1125^2 - 4(0.125)]^{1/2}}{2}$$

$$= -0.05625 \pm 0.349j$$

This shows that there are two complex poles, which are complex conjugates:

$$p_2 = -0.05625 + 0.349j$$

$$p_3 = -0.05625 - 0.349j$$

The factored form of the transform is, then,

$$V(s) = \frac{-1}{s(s + 0.05625 - 0.349j)(s + 0.05625 + 0.349j)} \qquad \blacksquare$$

These examples illustrate several important facts about the poles that appear in transforms:

1. The poles may be zero, real numbers, or complex numbers.
2. The real part of a pole is *always* negative.
3. When a complex pole $p = -x + jy$ appears, its complex conjugate $p = -x - jy$ also appears.

Computer. If the denominator has a power greater than 2, finding the roots can be very difficult. In these cases it is most advisable to use a computer program which finds the real and complex roots of polynomials. Without such help from the computer all but the most elementary problems in generalized network analysis become unmanageable.

EXAMPLE 4-14

Express the following Laplace transform in partially factored form as suitable for partial fraction expansion:

$$F(s) = \frac{-45s^3 + 24s - 66}{s^4 + 18s^3 + 137s^2 + 538s + 656}$$

SOLUTION

A computer program is used which finds the roots of polynomials. The program finds four roots and these roots are the poles;

$$p_1 = -2$$

$$p_2 = -4 + 5j$$

$$p_3 = -4 - 5j$$

$$p_4 = -8$$

This means that the transform can be written in the form of Eq. (4-19) as

$$F(s) = \frac{-45s^3 + 24s - 66}{(s + 2)(s + 4 - 5j)(s + 4 + 5j)(s + 8)} \qquad ■$$

4-3.2 Partial Fraction Expansion

Partial fraction expansion is essentially a mathematical procedure for doing the *opposite* of placing the sum and difference of fractions over a common denominator. The procedure takes a single fraction and expresses it as a sum and difference of "partial fractions." As applied to the problem of Laplace transforms it means that when a transform has been expressed in the form of Eq. (4-18) or (4-19), with the denominator factored, the transform can be written as a sum of fractions with the factors of the denominator as the individual denominators. Thus using partial fraction expansion, the transform given by Eq. (4-19) can be written in the form

$$F(s) = \frac{K_1}{s - p_1} + \frac{K_2}{s - p_2} + \cdots + \frac{K_n}{s - p_n} \qquad (4\text{-}20)$$

where $K_1, K_2, \ldots, K_n$ = real or complex constants. Now the point of all of this should start to become clear. By Table 4–1, once the transform has been expressed as a sum such as Eq. (4-20), each term can be *individually* inverted to find the time function. In other words, the time function of $F(s)$ will be the sum of the time functions of *each* of the terms in Eq. (4-20). Notice that the zeros do not appear explicitly in this expression, only the poles. The numerator of the transform serves only to find the values of the constants, K_i, while the poles define the time response of the transform.

Once a transform has been written in factored form, the functional form of the time dependence can be immediately written down. It is only the amplitudes of this dependence that will remain unknown. The amplitudes are determined by the constants in Eq. (4-20). In many cases the objective of solving some problem is simply to learn what *type* of time response is present in a network; the amplitudes of these signals is not important. In these cases the problem is finished when the transform has been expressed in the form of Eq. (4-20) or even Eq. (4-19), since the former is just an expansion into the factors of the latter. There are three classes of poles, with their respective time functions.

Class I: Real, Nonrepeated Poles.

This class of poles includes those with values of zero or negative real numbers, which occur only once. Table 4–1 shows that these poles contribute to the time function as follows:

1. A pole at zero, $p = 0$, means that partial fraction expansion will have a term proportional to $1/s$ which contributes a *dc level* to the time function.

2. A pole at some negative number, $p = -a$, means that the partial fraction expansion will have a term proportional to $1/(s + a)$ which contributes an *exponential decay* transient to the time function, as e^{-at}.

Class II: Real Repeated Poles.

This class of poles includes those with values of zero or negative real numbers, but which occur more than once in the transform. For example, the list of poles may show one at $p = -5$ and then continuing on down the list we see that a pole occurs at $p = -5$ again. This would be equivalent to an expression in the denominator such as $(s + 5)(s + 2)(s + 5) = (s + 5)^2 (s + 2)$. In this example the pole is repeated twice. Poles can be repeated any number of times, but in network analysis they are seldom repeated more than twice. These poles contribute as follows:

1. A repeated pole at zero means that the partial fraction expansion will contain terms proportional to $1/s$, $1/s^2$, $1/s^3$, all the way to the number of times it is repeated. If repeated only twice there will be terms $1/s$ and $1/s^2$. From Table 4–1 you can see that this means a contribution of a *dc level* and a linearly increasing or decreasing *ramp signal* proportional to t.

2. A repeated real negative number pole means that the partial fraction expansion will contain terms proportional to $1/(s + a)$, $1/(s + a)^2$, $1/(s + a)^3$, . . . all the way up to the number of times the pole is repeated in the transform. If repeated only twice, Table 4–1 shows that there will be time signals consisting of an *exponential decay*, e^{-at}, and a *ramped exponential decay*, te^{-at}.

Class III: Complex Poles. It has already been pointed out that when a complex pole appears in the transform, its complex conjugate also appears. The combination of a complex pole and its complex conjugate *always* result in a contribution of sine and/or cosine functions to the time signal. Table 4–1 shows that the denominator of sine and cosine functions has the form of $(s^2 + \omega^2)$ for a steady-state oscillation and $[(s + a)^2 + \omega^2]$ for a decaying oscillation. The fact that complex poles provide this type of transform can be seen from the following two types of complex poles and their contribution:

1. A purely complex pole has a zero real part, $p = j\omega$, and will always appear with its complex conjugate, $p^* = -j\omega$. The denominator of the transform will thus contain terms $(s + j\omega)$ and $(s - j\omega)$. But notice that

$$(s + j\omega)(s - j\omega) = s^2 + \omega^2$$

which is the form required for nondamped sine and/or cosine functions. Therefore, the presence of pure complex poles in the transform always results in *steady-state oscillation* consisting or sin (ωt) and/or cos (ωt).

2. A complex pole with a nonzero real part, $p = -a + j\omega$, will always occur with its complex conjugate, $p^* = -a - j\omega$. The denominator of the transform will thus contain terms $(s + a + j\omega)$ and $(s + a - j\omega)$. But notice if these are multiplied, then

$$(s + a + j\omega)(s + a - j\omega) = (s + a)^2 + \omega^2$$

which is the form for the damped sine and/or cosine functions. Therefore, the presence of complex poles of this type in the transform always results in *damped oscillation* consisting of e^{-at} sin (ωt) and/or e^{-at} cos (ωt).

Interpretation of the Transform. The descriptions above show that the functional dependence of the time signal which will result from some transform is known as soon as the poles have been found. Often, this is all that is required. The following examples show how such deductions are made.

EXAMPLE 4-15
Find the functional time dependence of the following transforms and express as partial fraction expansions.

(a) $I(s) = \dfrac{6s^2 + 2}{(s + 5)(s^2 + 16)}$

(b) $V(s) = \dfrac{3s^3 - 2s + 1}{s(s + 3)^2(s^2 + 6s + 8)}$

SOLUTION

(a) The quadratic term $(s^2 + 16)$ can be easily factored into $(s + 4j)(s - 4j)$. Thus the denominator has three poles: $p_1 = -5$, $p_2 = 4j$, and $p_3 = -4j$. The pole at -5 is a class I, real nonrepeated pole which therefore contributes e^{-5t} to the time dependence. The complex pair, $4j$ and $-4j$, are class III, purely complex, and therefore contribute $\sin(4t)$ and/or $\cos(4t)$. Partial fraction expansion according to Eq. (4-20) shows that $I(s)$ may be written

$$I(s) = \frac{K_1}{s + 5} + \frac{K_2}{s + 4j} + \frac{K_3}{s - 4j}$$

(b) It is clear that there is a pole at zero and a pole at -3 repeated twice. The poles of the quadratic must be determined by finding the roots:

$$\text{roots} = \frac{-6 \pm (36 - 32)^{1/2}}{2}$$

This gives two real nonrepeated poles at -4 and -2. So there is $p_1 = 0$, $p_2 = -3$ (twice), $p_3 = -2$, and $p_4 = -4$. The pole at zero is class I and contributes a dc level. The poles at -2 and -4 are class I and contribute e^{-2t} and e^{-4t}. The repeated pole at -3 is class II and contributes e^{-3t} and te^{-3t}. The partial fraction expansion is

$$V(s) = \frac{K_1}{s} + \frac{K_2}{s + 3} + \frac{K_3}{(s + 3)^2} + \frac{K_4}{s + 2} + \frac{K_5}{s + 4}$$

Notice that the pole at -3 contributes two terms to the partial fraction expansion, one for each power up to that for which it is repeated. ∎

EXAMPLE 4-16

What is the time dependence of the transform considered in Example 4-14?

SOLUTION

In Example 4-14 it was shown that the transform under consideration had poles at -2, -8, $-4 + 5j$, and $-4 - 5j$. The first two are class I and contribute e^{-2t} and e^{-8t}. The next two are a complex pair of class III which contribute

$$e^{-4t} \sin(5t) \quad \text{and/or} \quad e^{-4t} \cos(5t) \qquad ∎$$

4-3.3 Determination of the Amplitudes

Of course, there are many problems in network analysis where it is not sufficient merely to know the functional time dependence of a transform. In these cases it is necessary for the solution to give the actual amplitudes of the individual terms. This means that the values of the K_i in Eq. (4-20) must be found. It is convenient to consider how the amplitudes are found in terms of the three classes of poles introduced in the preceding section.

Class I: Real Nonrepeated Poles. The amplitudes for the real nonrepeated poles in the transform can be found from a very simple equation using the Laplace transform in the form expressed by Eq. (4-19).

$$K_i = [(s - p_i)F(s)]_{s=p_i} \tag{4-21}$$

where K_i = amplitude for real, nonrepeated pole, p_i. It is helpful to understand why Eq. (4-21) works for finding the value of the amplitude. Consider for a moment the result of multiplying $F(s)$ by the term $(s - p_i)$. Since p_i is one of the poles of $F(s)$, one of the factors in the denominator *will be* the term $(s - p_i)$ itself! This means that multiplying $F(s)$ by the factor will simply cancel it from the transform:

$$(s - p_i)F(s) = (s - p_i) \frac{a_m s^m + \cdots + a_1 s + a_0}{(s - p_1) \cdots (s - p_i) \cdots (s - p_n)}$$

Of course, this really must happen since if the pole were not canceled, substitution of $s = p_i$ would cause the transform to go to infinity. Now to see how this picks out the value of K_i consider the effect of multiplying $(s - p_i)$ times the transform in the expanded form of Eq. (4-20):

$$(s - p_i)F(s) = (s - p_i) \frac{K_1}{s - p_1} + \cdots + K_i + \cdots + (s - p_i) \frac{K_n}{s - p_n}$$

The K_i term is alone because the $(s - p_i)$ canceled in that term and that term only. Now, in this last equation, when $s = p_i$ every term will become zero except the constant amplitude, K_i. Thus combining the two results above leads to the relationship of Eq. (4-21).

EXAMPLE 4-17
Find the time function of the following Laplace transform:

$$F(s) = \frac{3s + 2}{s^2 + 8s + 12}$$

SOLUTION
The roots are found from the quadratic formula

$$\text{roots} = \frac{-8 \pm (8^2 - 48)^{1/2}}{2} = -4 \pm 2$$

So the roots and therefore the poles are $p_1 = -6$ and $p_2 = -2$. The transform can now be written in the form of Eq. (4-19) and expanded as Eq. (4-20):

$$F(s) = \frac{3s + 2}{(s + 6)(s + 2)} = \frac{K_2}{s + 6} + \frac{K_2}{s + 2}$$

Both of the poles are real, nonrepeated so that the values of the constant amplitudes can be found using Eq. (4-21).

$$K_1 = (s + 6)F(s) \bigg]_{s=-6} = (s + 6)\frac{3s + 2}{(s + 6)(s + 2)}\bigg]_{s=-6}$$

$$= \frac{3(-6) + 2}{-6 + 2} = 4$$

$$K_2 = (s + 2)F(s) \bigg]_{s=-2} = (s + 2)\frac{3s + 2}{(s + 6)(s + 2)}\bigg]_{s=-2}$$

$$= \frac{3(-2) + 2}{-2 + 6} = -1$$

The complete transform can now be written as

$$F(s) = \frac{4}{s + 6} - \frac{1}{s + 2}$$

You can prove that this is the same transform by putting the terms over a common denominator. The time function can now be found by using Table 4–1 on each term of the transform,

$$f(t) = 4e^{-6t} - e^{-2t} \qquad \blacksquare$$

Class II: Real Repeated Poles. The procedure and equation given in the preceding section will not work for finding the constant amplitudes of repeated poles. In network analysis poles will usually not be repeated more than twice, but a general formula will be given which finds the constants for any number of repeats. To set up the background for the amplitude equation, let us assume that some pole, p_i, is repeated q times in a transform. The transform will then have the form

$$F(s) = \frac{A(s)}{(s - p_1) \cdots (s - p_i)^q \cdots (s - p_n)} \qquad (4\text{-}22)$$

Note that the pole factor appears to the qth power since it is repeated q times. When the partial fraction expansion is performed it turns out there will be a term in the expansion for every power of the repeated pole. Thus there will be q terms for the pole p_i. The terms will be to each power of the factor from 1 to q and each will have a constant amplitude. Then the expanded form of the transform is

$$F(s) = \frac{K_1}{s - p_1} + \cdots + \frac{K_{i1}}{s - p_i} + \frac{K_{i2}}{(s - p_i)^2}$$

$$+ \cdots + \frac{K_{iq}}{(s - p_i)^q} + \cdots + \frac{K_n}{s - p_n} \tag{4-23}$$

where

$$K_1 \cdots K_n = \text{amplitudes of nonrepeated poles}$$

$$K_{i1} K_{i2} \cdots K_{iq} = \text{amplitudes of repeated pole}$$

Note the q terms that the repeated pole contributes to the expansion. Now, the equation that allows one to find the amplitudes of the repeated poles can be written in the form

$$K_{ij} = \frac{1}{(q - j)!} \frac{d^{q-j}}{ds^{q-j}} \left[(s - p_i)^q F(s) \right]_{s=p_i} \tag{4-24}$$

where

$$K_{ij} = \text{the } j\text{th constant from the } q \text{ amplitudes of } p_i; \text{ it is the}$$

$$\text{amplitude for } (s - p_i)^j$$

$$j = 0, 1, 2, \ldots, q$$

Equation (4-24) allows for finding all q constants for the contributions of the repeated pole to the transform. The following special notes should be made about use of this equation:

1. The term $(q - j)!$ means *factorial*, which, you recall, is defined by a product such that $N! = 1 \cdot 2 \cdot 3 \cdots N$. For the constant with $j = q$, we have $(q - q)! = 0!$, which is unity, $0! = 1$.
2. The derivative is taken $(q - j)$ times prior to substitution of the pole for s. When $j = q$ the derivative is taken $(q - q) = 0$ times. This means do not differentiate.

EXAMPLE 4-18

Find the time function of the transform

$$F(s) = \frac{3s + 1}{s(s + 2)(s + 3)^2}$$

SOLUTION

You can see that there are real nonrepeated poles at 0 and −2 and a twice-repeated pole at −3. From Eq. (4-23) the expansion will be

$$F(s) = \frac{K_1}{s} + \frac{K_2}{s + 2} + \frac{K_{31}}{s + 3} + \frac{K_{32}}{(s + 3)^2}$$

K_1 and K_2 are found from the real nonrepeated pole amplitude formula, Eq. (4-21).

$$K_1 = sF(s) \bigg]_{s=0} = \tfrac{1}{18}$$

$$K_2 = (s + 2)F(s) \bigg]_{s=-2} = \tfrac{5}{2}$$

The repeated pole amplitudes are found from Eq. (4-24). For K_{31} we have $q = 2$ and $j = 1$; thus

$$K_{31} = \frac{1}{(2 - 1)!} \frac{d^{2-1}}{ds^{2-1}} \left[(s + 3)^2 \frac{3s + 1}{s(s + 2)(s + 3)^2} \right]_{s=-3}$$

$$= \frac{d}{ds} \left[\frac{3s + 1}{s^2 + 2s} \right] = \frac{(s^2 + 2s)(3) - (3s + 1)(2s + 2)}{(s^2 + 2s)^2} \bigg]_{s=-3}$$

$$= -\tfrac{23}{9}$$

$$K_{32} = \frac{1}{(2 - 2)!} \frac{d^{2-2}}{ds^{2-2}} \left[\frac{3s + 1}{s^2 + 2s} \right] = \frac{3s + 1}{s^2 + 2s} \bigg]_{s=-3}$$

$$= -\tfrac{8}{3}$$

The transform now has the form

$$F(s) = \frac{1}{18s} + \frac{5}{2} \frac{1}{s + 2} - \frac{23}{9} \frac{1}{s + 3} - \frac{8}{3} \frac{1}{(s + 3)^2}$$

Now the time function can be written using Table 4–1:

$$f(t) = \tfrac{1}{18} + \tfrac{5}{2}e^{-2t} - \tfrac{23}{9}e^{-3t} - \tfrac{8}{3}te^{-3t}$$ ∎

Class III: Complex Poles. When a complex pole appears with its complex conjugate in a Laplace transform it always means that the time function will have sine and/or cosine terms. The problem of amplitudes becomes one of determination of the amplitude of the sine/cosine functions. Two methods for finding these amplitudes will be presented.

Method of Complex Amplitudes. The first method is based on the fact that Eq. (4-21) will work equally well for complex poles as for real nonrepeated poles. The problem is that the resulting amplitudes are complex numbers and are not, directly, the amplitude of the sine or cosine functions. To see how all this comes together, consider a transform which, for convenience, contains only a complex pole pair, $p = -a + j\omega$ and $p^* = -a - j\omega$. You will recall that the real part of such a pole was earlier identified to be the damping and is always a negative number. The imaginary part was identified to be the angular frequency. The transform will be

$$F(s) = \frac{A(s)}{(s + a + j\omega)(s + a - j\omega)} \qquad (4\text{-}25)$$

According to Eq. (4-20), the transform of Eq. (4-25) can be expanded as

$$F(s) = \frac{K_1}{s + a + j\omega} + \frac{K_2}{s + a - j\omega} \qquad (4\text{-}26)$$

The two constant amplitudes in Eq. (4-26) will turn out to be complex numbers *and* will be complex conjugates, $K_2 = (K_1)^*$. This means that if one of the constants is found, the other is given by same numbers with opposite sign for the complex part. It is possible to relate the value of K_1 to the amplitudes of the sine and cosine functions of time. To do this let us write the amplitude in terms of real and imaginary parts, $K_1 = R + jI$. Then Eq. (4-26) becomes

$$F(s) = \frac{R + jI}{s + a + j\omega} + \frac{R - jI}{s + a - j\omega}$$

This equation is now put over a common denominator:

$$F(s) = \frac{(R + jI)(s + a - j\omega) + (R - jI)(s + a + j\omega)}{(s + a + j\omega)(s + a - j\omega)}$$

When the numerator is multiplied out, the imaginary parts cancel, leaving then only real terms. The denominator has been shown before to be $(s + a)^2 + \omega^2$, which is the denominator form for a sine/cosine function with damping. All of this allows the equation above to be written in the form

$$F(s) = 2R \frac{s + a}{(s + a)^2 + \omega^2} + 2I \frac{\omega}{(s + a)^2 + \omega^2} \qquad (4\text{-}27)$$

But the transform expressions in Eq. (4-27) can be recognized from Table 4–1 to be the terms for the cosine and sine functions. Thus the time function is

$$f(t) = 2Re^{-at} \cos (\omega t) + 2Ie^{-at} \sin (\omega t) \qquad (4\text{-}28)$$

In summary, then, when a complex pole pair occurs in a transform, perform the partial fraction expansion treating these terms as nonrepeated poles. Find the complex amplitude for the pole at $-a - j\omega$ using the nonrepeated pole formula, Eq. (4-20). Then the part of the transform for the complex pair can be written in the form of Eq. (4-27) with the amplitude of the cosine term being twice the real part and the amplitude of the sine term being twice the imaginary part. This works for cases with no damping by simply using $a = 0$ in the relations above.

EXAMPLE 4-19

Find the time function of the following Laplace transform:

$$V(s) = \frac{6s + 10}{s(s^2 + 4s + 20)}$$

SOLUTION

It is easy to show that the quadratic has roots of $-2 - 4j$ and $-2 + 4j$, which means that this term is a complex pole pair. Then, of course, there is a pole at zero. The transform can thus be expanded as

$$V(s) = \frac{K_1}{s} + \frac{K_2}{s + 2 + 4j} + \frac{(K_2)^*}{s + 2 - 4j}$$

where it has been noted that the constants are complex conjugates. The constant K_1 is found as

$$K_1 = sF(s) \Big]_{s=0} = 0.5$$

The constant K_2 is found from Eq. (4-20) as

$$K_2 = (s + 2 + 4j) \left. \frac{6s + 10}{s(s + 2 + 4j)(s + 2 - 4j)} \right]_{s = -2 - 4j}$$

$$= \frac{6(-2 - 4j) + 10}{(-2 - 4j)(-2 - 4j + 2 - 4j)} = \frac{-2 - 24j}{-32 + 16j}$$

The last expression can be converted to a single complex number by multiplying top and bottom by $-32 - 16j$. The result is $K_2 = -0.25 + 0.625j$. So, considering Eq. (4-27), with $R = -0.25$, $I = 0.625$ and

$$V(s) = \frac{0.5}{s} - 0.5 \frac{s + 2}{(s + 2)^2 + 16} + 1.25 \frac{4}{(s + 2)^2 + 16}$$

The time function is found directly from Table 4-1:

$$v(t) = 0.5 - 0.5e^{-2t} \cos (4t) + 1.25e^{-2t} \sin (4t) \qquad \blacksquare$$

Method of Matching. A second method of finding the amplitudes of complex poles associated with a Laplace transform is based on the fact that the original transform expressed as Eq. (4-19) must equal the expanded transform of Eq. (4-20). The basic procedure is to put the expression of Eq. (4-20) over a common denominator, which will be the original denominator of Eq. (4-19), and then find values of the constants to make the numerators equal. This method will actually find the values of all the constants in the partial fraction expansion, but for real poles the formulas are probably easier.

The idea in this approach is to use the knowledge that complex pole pairs in the transform *must* give rise to sine and/or cosine terms in the time function. Thus, whenever a complex pole pair appears, the partial fraction expansion is *automatically* set down with the transforms of the sine and cosine functions *instead* of complex poles. Thus, using Eq. (4-25) for illustration, when the complex pole pair is noted in this transform, the transform is simply written

$$F(s) = \frac{A\omega}{(s + a)^2 + \omega^2} + \frac{B(s + a)}{(s + a)^2 + \omega^2} \qquad (4\text{-}29)$$

where

$$A = \text{amplitude for the sine term}$$

$$B = \text{amplitude for the cosine term}$$

From reference to Table 4–1, you can see that the appropriate transforms for the sine and cosine functions have been written down with an unknown constant. This was done because complex pole pairs *mean* that such sine and cosine terms will be present. Now, the constant amplitudes, A and B, are found by putting the transform over a common denominator and making the numerator match the numerator of the original transform. The following example illustrates this for a previous problem so that a comparison can be made between the two methods.

EXAMPLE 4-20

Find the time function of the transform of Example 4-19 using the method of matching.

SOLUTION

The first step of the problem is to find the poles, which was done in Example 4-19 and found to be a pole at zero and a complex pair, $-2 - 4j$ and $-2 + 4j$. By the method of substitution, the transform would then directly be written

$$F(s) = \frac{K_1}{s} + \frac{4A}{(s + 2)^2 + 16} + \frac{B(s + 2)}{(s + 2)^2 + 16}$$

Now this transform is put over a common denominator with $(s + 2)^2 + 16 = s^2 + 4s + 20$:

$$F(s) = \frac{K_1[(s + 2)^2 + 16] + s[4A + B(s + 2)]}{s(s^2 + 4s + 20)}$$

The numerator is now expressed as a polynomial:

$$F(s) = \frac{[K_1 + B]s^2 + [4K_1 + 4A + 2B]s + 20K_1}{s(s^2 + 4s + 20)}$$

This must still equal the original transform. Since the denominators are already equal, the numerators must be equal:

$$6s + 10 = [K_1 + B]s^2 + [4K_1 + 4A + 2B]s + 20K_1$$

This must be true for any s, which can only be satisfied if the coefficients of like powers of s are equal, giving the equations

$$0 = K_1 + B \qquad \text{for the coefficient of } s^2$$

$$6 = 4K_1 + 4A + 2B \qquad \text{for the coefficient of } s^1$$

$$10 = 20K_1 \qquad \text{for the coefficient of } s^0$$

The last gives $K_1 = 10/20 = 0.5$. The first shows that $B = -K_1 = -0.5$. Finally the second equation gives $A = (6 - 2 + 1)/4 = 1.25$. This gives a time function of

$$f(t) = 0.5 - 0.5e^{-2t} \cos(4t) + 1.25e^{-2t} \sin(4t)$$

which is, of course, the same as the result of Example 4-19. ∎

The same procedures are followed for any problem solved by matching. An advantage is that this method provides a check on the accuracy of constants found by the formulas.

4-4

INTERPRETATIONS AND APPLICATIONS

Laplace transforms are used extensively in many fields of study for many different purposes. Most often, the use stems from a desire to solve sets of differential equations which have been derived to describe some process. In this chapter the Laplace transform has been shown to be a valuable tool for solving general electrical networks. The purpose of this section is twofold. First, it will be shown that much information about the time behavior of network variables can be obtained from the Laplace transforms without finding the time functions explicitly. Second, an application of Laplace transforms to a sample network problem will be given to illustrate the procedures to be followed in any problem.

4-4.1 Interpretation of the Laplace Transform

In many cases it is possible to answer questions about network behavior by examination of the Laplace transform, thereby eliminating the labor of inverting the transform to find the time function. There are a number of features of the transform that provide for this capability.

Time Response. As was pointed out earlier in this chapter, once the denominator of a transform has been evaluated the form of the time dependence is known, including decay times and frequencies of oscillation, if any. The poles define the time response. Often this is all that is desired.

EXAMPLE 4-21
Analysis of a network resulted in the following Laplace transform of the output voltage. From the transform explain what the output signal form is after all

transients have died out. How long after $t = 0$ before the longest-lived transient has died?

$$V_{out}(s) = \frac{-33s^3 + 244s^2 - 66}{s^5 + 6s^4 + 24s^3 + 96s^2 + 128s}$$

SOLUTION

The first step is to find the poles. This is done using a computer program which finds the roots of the polynomial in the denominator. The results are

$$p_1 = 0 \quad \rightarrow \quad \text{dc level}$$

$$p_2 = -2 \quad \rightarrow \quad e^{-2t}$$

$$p_3 = -4 \quad \rightarrow \quad e^{-4t}$$

$$p_4 = 4j \quad \rightarrow \quad \text{the complex pair give sin } (4t) \text{ and/or cos } (4t)$$

$$p_5 = -4j$$

The two exponentials are transients, so after a certain length of time they will die away. What will be left is a dc level with an oscillation at a frequency of 4 rad/s or $f = \omega/2\pi = 0.64$ Hz. The time for transients to decay away is given approximately by five times the time constant of the exponential term. One time constant is $\frac{1}{2}$ s and the other $\frac{1}{4}$ s. Clearly, the first is the longest lived and will die away in $\frac{5}{2}$ or 2.5 s. ∎

Initial and Final Value Theorems.   It is possible in most cases to deduce the initial and final values of a time function from its Laplace transform, without actually inverting the transform. This is often used as a check against the possibility of errors in a calculation. For example, the initial value of a network variable as deduced from its transform must equal the value of the variable deduced from initial conditions.

1. Initial Value Theorem.   The value of a time function at $t = 0$, $f(0)$, is given by the limit of $sF(s)$ as s goes to infinity, where $F(s)$ is the transform of $f(t)$:

$$f(0) = \lim_{s \to \infty} [sF(s)] \tag{4-30}$$

2. Final Value Theorem.   The value of a time function $f(t)$ as t goes to infinity is given by the limit of $sF(s)$ as s goes to zero, provided that the transform $F(s)$ contains no purely complex poles, $p = j\omega$.

$$\lim_{t \to \infty} [f(t)] = \lim_{s \to 0} [sF(s)] \tag{4-31}$$

The reason for the restriction on real poles in the second theorem is due to the fact that a purely complex pole will mean a steady-state oscillation. Of course, a steady-state oscillation will not have a *final* value since it oscillates between two values.

EXAMPLE 4-22

Apply the initial and final value theorems to the transform of Example 4-14.

SOLUTION

In Example 4-14 it was shown that there were no purely complex poles, so both limiting evaluations can be performed. The initial value theorem gives

$$f(0) = \lim_{s \to \infty} s \frac{-45s^3 + 24s - 66}{s^4 + 18s^3 + 137s^2 + 538s + 656}$$

To evaluate a limit to infinity, we try to get each term to be in the form of $1/s$ since these will go to zero as s goes to infinity. This can be done by dividing every term by s^4.

$$f(0) = \lim_{s \to \infty} \frac{-45 + 24/s^2 - 66/s^3}{1 + 18/s + 137/2^2 + 538/s^3 + 656/s^4}$$

Now every term in $1/s$ to any power will go to zero, leaving

$$f(0) = \lim_{s \to \infty} [sF(s)] = -45$$

For the final value theorem the lim of s going to zero is simply passed:

$$\lim_{t \to \infty} f(t) = \lim_{s \to 0} \frac{-45s^4 + 24s^2 - 66s}{s^4 + 18s^3 + 137s^2 + 538s + 656} = 0$$

so that the final value will be zero, $f(\infty) = 0$. ■

4-4.2 Application of Laplace Transforms to Network Analysis

The global objective of introducing Laplace transforms was to provide a tool for the solution of networks with capacitors, inductors, and transformers. In this section a single network problem will be solved using Laplace transforms to illustrate the approach and procedure. In general, the solution of a network can be found through the following steps:

1. Determine the initial conditions.
2. Derive the integrodifferential equations for the network using KCL or KVL.

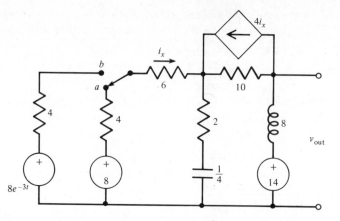

FIGURE 4–6. Network to illustrate the steps involved in using Laplace transforms to solve a problem.

3. Find the Laplace transforms of these equations.
4. Solve for the Laplace transform of the desired unknown or unknowns.
5. Invert the transform to determine the time function.

In one form or another these same steps are repeated whenever a network is solved using Laplace transforms.

The Problem. For the network of Fig. 4–6 the switch moves from position *a* to position *b* at $t = 0$. Notice that this inserts an input signal to the network which has the form of Fig. 4–7. Such a signal is typical of the trailing edge of a square wave or square pulse. The problem is to find the solution for $v_{out}(t)$. This will show how such a trailing edge shows up in the output signal.

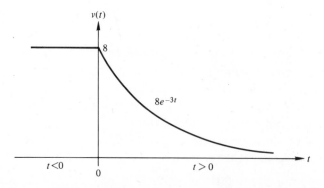

FIGURE 4–7. Effective form of the input signal for the network in Fig. 4–6.

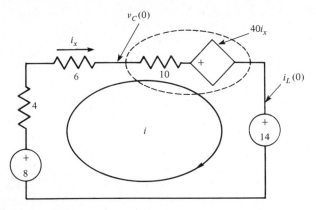

FIGURE 4–8. The initial conditions for the network of Fig. 4–6 are found from this dc, resistive network.

Initial Conditions. The initial conditions are defined by finding the voltage on capacitors and currents through inductors in the initial state, $t = 0^-$. This state is defined by switch position a when the configuration of the network is as shown in Fig. 4–8. This problem is solved using dc mesh analysis. Note that the location of the inductor and capacitor have been marked in the network. The current source has been transformed into a voltage source. The single-loop mesh equation for the current becomes

$$(20)i = 8 - 4i_x - 14 = -6 - 40i_x$$

where

$$i = \text{loop current}$$

$$i_x = \text{controlled source current} = i \text{ in this case}$$

The current can thus be found as

$$60i = -6$$

$$i = -0.1 \text{ A}$$

This is also the current that flows through the inductor, so the initial condition on the inductor is

$$i_{8H}(0) = -0.1 \text{ A} \tag{4-32}$$

where the negative notes that the current is counterclockwise. The capacitor voltage can now be found as the voltage between the indicated locations of the network. This is found as the 8-V source combined with the voltage dropped across the 4- and 6-Ω resistors,

$$V_{0.25F}(0) = 8 + 10(0.1) = 9 \text{ V} \tag{4-33}$$

where the voltage drop adds to the 8 V because the counterclockwise current produces a voltage drop positive on the side of the resistors from which the current enters.

Network Equations. The integrodifferential equations are found by moving the switch to position b and applying KVL or KCL to the resulting network. The elements tend to be in series, so KVL will be used. The network to be solved is shown in Fig. 4–9, with loop currents defined and the current source transformed into a voltage source for convenience in using KVL. With the aid of the $i–v$ relations of Table 3–1, the resulting equations for the two loops become

$$12i_1 + 4\int_0^t i_1(\tau)\, d\tau - 2i_2 - 4\int_0^t i_2(\tau)\, d\tau = 8e^{-3t} - 9$$

$$- 2i_1 + 40i_x - 4\int_0^t i_1(\tau)\, d\tau + 12i_2 + 4 \tag{4-34}$$

$$\times \int_0^t i_2(\tau)\, d\tau + 8\frac{di_2}{dt} = -5$$

but it is clear that $i_x = i_1$, so the last equation becomes

$$38i - 4\int_0^t i_1(\tau)\, d\tau + 12i_2 + 4\int_0^t i_2(\tau)\, d\tau + 8\frac{di_2}{dt} = -5 \tag{4-35}$$

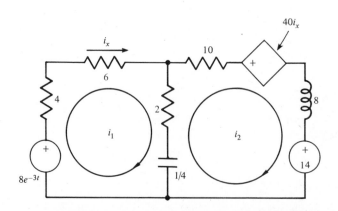

FIGURE 4-9. Network to be solved for the output voltage in Fig. 4—6.

These two simultaneous equations for the currents in the loop will solve for the required unknown.

Laplace Transformation. The next step is to find the Laplace transform of the equations given by Eqs. (4-34) and 4-35). This is done by using Tables 4–1 and 4–2 to replace all functions and operations in the equations by the corresponding Laplace transform. It will be necessary to include the initial condition of the inductor because of the appearance of the derivative of the current, due to the inductor. The capacitor initial condition has already been included in the integral term of this element. The resulting equations are

$$\left(12 + \frac{4}{s}\right)I_1(s) - \left(2 + \frac{4}{s}\right)I_2(s) = \frac{8}{s + 3} - \frac{9}{s}$$

$$\left(38 - \frac{4}{s}\right)I_1(s) + \left(12 + \frac{4}{s} + 8s\right)I_2(s) = -\frac{5}{s} - 0.8$$

This set of equations can be placed in our familiar *matrix* form as follows:

$$\begin{pmatrix} 12 + \dfrac{4}{s} & -2 - \dfrac{4}{s} \\ 38 - \dfrac{4}{s} & 12 + \dfrac{4}{s} + 8s \end{pmatrix} \begin{pmatrix} I_1(s) \\ I_2(s) \end{pmatrix} = \begin{pmatrix} \dfrac{8}{s + 3} - \dfrac{9}{s} \\ -\dfrac{5}{s} - 0.8 \end{pmatrix} \qquad (4\text{-}36)$$

The matrix equation given by Eq. (4-36) represents the set of equations necessary to find a complete solution to the network of Fig. 4–6. Cramer's rule and determinant evaluation, including zero generation and Laplace expansion, can be used to evaluate the unknowns in the matrix equation.

Solving for the Unknown. In this particular problem the required answer is the voltage across the output, which is across the inductor and 14-V source. Using the known relation between inductor current and voltage, the output voltage can be written

$$v_{\text{out}}(t) = 14 + 8\frac{di_2}{dt}$$

Taking Laplace transforms of this equation gives

$$V_{\text{out}}(s) = \frac{14}{s} + 8sI_2(s) + 0.8 \qquad (4\text{-}37)$$

Equation (4-37) shows that a solution for $I_2(s)$ is required in order to solve for the output voltage. Notice the 0.8 term, which comes from the initial condition of the inductor through the transform of the derivative operation. Now it will be necessary to solve Eq. (4-36) for $I_2(s)$ and substitute this into Eq. (4-37) to find the Laplace transform of the required answer. Using Cramer's rule, the solution for $I_2(s)$ can be expressed as

$$I_2(s) = \cfrac{\begin{vmatrix} 12 + \dfrac{4}{s} & \dfrac{8}{s+3} - \dfrac{9}{s} \\[3mm] 38 - \dfrac{4}{s} & -\dfrac{5}{s} - 0.8 \end{vmatrix}}{\begin{vmatrix} 12 + \dfrac{4}{s} & -2 - \dfrac{4}{s} \\[3mm] 38 - \dfrac{4}{s} & 12 + \dfrac{4}{s} + 8s \end{vmatrix}}$$

When multiplied out using the normal determinant evaluation rules this becomes

$$I_2(s) = \cfrac{\left(12 + \dfrac{4}{s}\right)\left(-\dfrac{5}{s} - 0.8\right) - \left(38 - \dfrac{4}{s}\right)\left(\dfrac{8}{s+3} - \dfrac{9}{s}\right)}{\left(12 + \dfrac{4}{s}\right)\left(12 + \dfrac{4}{s} + 8s\right) + \left(2 + \dfrac{4}{s}\right)\left(38 - \dfrac{4}{s}\right)} \tag{4-38}$$

The last equation is the transform of the current required for insertion into Eq. (4-37) for the output voltage.

Inversion of the Transform. It will now be necessary to find the time solution for the output voltage by inserting $I_2(s)$ into Eq. (4-37) and taking the inverse transform. As a step toward doing this it will be first necessary to perform partial fraction expansion of Eq. (4-38). The first step in this process is to place the transform in *good form* by expressing it as the ratio of two polynomials in s. This is done by multiplying out the numerator and denominator and then multiplying each by what ever terms will eliminate inverse s and $s + 3$ terms. The result of this, including the $8s$ multiplier of Eq. (4-37), becomes

$$8sI_2(s) = -\frac{0.8s^3 + 4.5s^2 - 67.7s + 14}{(s + 3)(s^2 + 2.625s + 2.5)} \tag{4-39}$$

The factor $(s + 3)$ has not been multiplied into the denominator since one of the objectives is to factor the denominator. This term is clearly already one of the factors representing a pole at $p = -3$ and it would be going backward to multiply it into the quadratic. However, there is another problem. Remember that to perform partial fraction expansion, a power test must be performed. This transform does not satisfy this test since both numerator and denominator are of the same highest power in s, namely s^3. Thus it is necessary to divide the polynomials in Eq. (4-39). The result gives the expression

$$8sI_2(s) = -0.8 + \frac{76s - 8}{(s + 3)(s^2 + 2.625s + 2.5)} \tag{4-40}$$

When this expression is substituted into Eq. (4-37) for the output voltage transform, the 0.8 term cancels and the resulting expression is

$$V_{out}(s) = \frac{14}{s} + \frac{76s - 8}{(s + 3)(s^2 + 2.625s + 2.5)} \tag{4-41}$$

The poles of the quadratic are given by the roots of this expression:

$$\text{roots} = -\frac{2.625}{2} \pm \frac{(2.625^2 - 10)^{1/2}}{2} \approx -1.3 \pm 0.9j$$

Thus the complex pair show that there will be a transient oscillation and the pole at -3 shows that there is a simple exponential decay. This means that the expression for the output voltage has the expanded form of

$$V_{out}(s) = \frac{14}{s} + \frac{K_1}{s + 3} + \frac{0.9A}{(s + 1.3)^2 + 0.81} + \frac{(s + 1.3)B}{(s + 1.3)^2 + 0.81} \tag{4-42}$$

Evaluation of the amplitudes can be carried out with either of the methods outlined earlier in this chapter. Remember that for this evaluation only the second term of the expression in Eq. (4-41) is used. Using the formula for the real, nonrepeated pole gives

$$K_1 = (s + 3) \frac{76s - 8}{(s + 3)(s^2 + 2.6s + 2.5)} \Bigg]_{s=-3}$$

$$= \frac{-228 - 8}{9 - 7.8 + 2.5} = -63.8$$

To solve for A and B in Eq. (4-42), the method of matching will be used. This will also give a check on the answer for K_1. When the other terms of Eq. (4-42) are multiplied out and placed over a common denominator, the numerator must be equal to the expression $76s - 8$. This gives the following equality:

$$76s - 8 = K_1(s^2 + 2.6s + 2.5) + 0.9A(s + 3) + B(s + 1.3)(s + 3)$$

When the expression on the right is multiplied out and rearranged the result is

$$76s - 8 = (K_1 + B)s^2 + (2.6K_1 + 0.9A + 4.3B)$$
$$+ (2.5K_1 + 2.7A + 3.9B)$$

This gives three equations in three unknowns:

$$0 = K_1 + B$$
$$76 = 2.6K_1 + 0.9A + 4.3B$$
$$-8 = 2.5K_1 + 2.7A + 3.9B$$

These can be quickly solved by Cramer's rule to give the solutions

$$K_1 = -63.8$$
$$B = 63.8$$
$$A = -36$$

This now allows the complete transform for the output voltage to be written, with all amplitudes determined:

$$V_{out}(s) = \frac{14}{s} - \frac{63.8}{s+3} + \frac{63.8(s+1.3)}{(s+1.3)^2 + 0.81} - \frac{36(0.9)}{(s+1.3)^2 + 0.81} \quad (4\text{-}43)$$

This transform can be easily inverted using the table to give the time function

$$v_{out}(t) = 14 - 63.8e^{-3t} + 63.8e^{-1.3t}\cos(0.9t)$$
$$- 36e^{-1.3t}\sin(0.9t) \quad (4\text{-}44)$$

The result, given in Eq. (4-44), is plotted in Fig. 4–10.

4-4.3 Computer Applications

The use of the computer has already been seen in finding the roots of denominator polynomials in order to find the transform poles. Another important use of the computer employs programs that perform partial fraction expansion. Usually, such programs require input of the orders and coefficients of the numerator and denominator polynomials. From this the partial fraction expansion is performed and the poles and amplitudes are outputted. This then allows immediate construction of the time function. Of course, the other common and very important application of the computer in network analysis is for the plotting of the results of analysis.

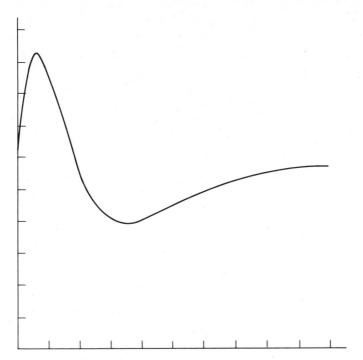

FIGURE 4–10. Plot of the output voltage found for the network of Fig. 4–6.

The next chapter illustrates the application of computer program packages which find the time function from Laplace transforms.

SUMMARY

This chapter introduced the Laplace transform method of solving problems in network analysis. The specific tasks considered in this chapter included the following major items:

1. A definition was given of the Laplace transform of a time function in terms of an integration of the time function.
2. It was shown that operations in time, such as integration and differentiation, can be transformed as well as functions of time.
3. Transformations of time functions can be made by the integral definition or by use of *tables* of standard time functions and their corresponding Laplace transform.
4. The tables can be used to convert a time function into a Laplace transform or a Laplace transform into its equivalent time function.
5. A table can also be used to transform operations in time into equivalent operations in the s-space of Laplace transforms.
6. In many cases a Laplace transform of some time function will appear in a

form that makes identification from the table very difficult. It is necessary to simplify the transform so that the table can be used.

7. Partial fraction expansion is a method by which a complex Laplace transform can be broken down into simpler parts which can be identified from the tables. The method involves a number of steps:

   (a) Express the transform as the ratio of two polynomials.

   (b) Verify that the highest power of s in the denominator is greater than the numerator, or divide until the remainder satisfies this test.

   (c) Find the roots of the denominator, which are the poles.

   (d) Find the partial fraction expansion by writing the transform as the sum of individual pole fractions.

8. Now that the transform has been expressed as the sum of simple transforms, the table can be used to find the time function. Each pole denotes a certain type of time dependence.

   (a) A pole at zero is a constant, dc, term.

   (b) A real pole is always negative and denotes exponential decay.

   (c) A complex pole always appears with its complex conjugate and always denotes oscillation expressed as sine and cosine time functions.

9. The constants resulting from the partial fraction expansion are evaluated by formulas or numerator matching methods.

PROBLEMS

4-1   Use the integral definition to find the Laplace transform of the time function $f(t) = 5t$.

4-2   Use the table of time function and Laplace transform pairs to find the transforms of the following functions of time.

(a) $f(t) = 5 + 2e^{-3t}$ (b) $f(t) = 6te^{-5t}$

(c) $f(t) = 4(1 - e^{-6t})$ (d) $f(t) = 7e^{-3t}[1 + \cos(4t)]$

(e) $f(t) = 3e^{-5t}\sin(2t) - 7e^{-5t}\cos(2t)$

4-3   Use the time function/Laplace transform table to find the time functions of the following transforms.

(a) $F(s) = 4 + \dfrac{4}{s}$ (b) $F(s) = -\dfrac{6}{s} + \dfrac{3}{s + 7}$

(c) $F(s) = \dfrac{3s}{s^2 + 9} + \dfrac{27}{s^2 + 9}$ (d) $F(s) = \dfrac{6s + 12}{s^2 + 4s + 20}$

(e) $F(s) = \dfrac{2}{(s + 4)^2 + 25}$

4–4   Use the table of time operations and Laplace transform operations to transform the following relations.

(a) $5 + 8f(t)$ (b) $7f(t) - 8\dfrac{df}{dt}$; $f(0) = -3$

(c) $3\dfrac{df}{dt} + \dfrac{1}{5}\displaystyle\int_0^t f(\tau)\, d\tau - 7f(t)$; $f(0) = 4$

4–5   Given a function $f(t) = 3t - 6\sin(4t)$, show how the derivative can be found using Laplace transforms.

4–6   Given the equation

$$5f(t) + 45\int_0^t f(\tau)\, d\tau - 15 = 0$$

if $f(0) = 0$, find $f(t)$ using Laplace transforms.

4–7   Find the Laplace transform equations of the integrodifferential equations found in (a) Example 3-4, (b) Example 3-6, and (c) Example 3-8.

4–8   Use Cramer's rule to find the Laplace transform of node 3 voltage of Example 3-8. Express this transform in good form, as the ratio of two polynomials in s.

4–9   Specify the form of the time dependence of the transform

$$F(s) = \frac{6 - 2/s - 1/s^2}{s(s + 6)^2(1 + 1/s)(s + 4/s)}$$

4–10   Find the time functions of the following transforms.

(a) $F(s) = \dfrac{15 + 10/s}{(1 + 3/s)(5s + 5)}$ (b) $F(s) = \dfrac{6s^2 - 8/s}{2s^2 + 10s + 8}$

(c) $F(s) = \dfrac{2s^2 - 5s + 8}{(s + 2)(s^2 + 6s + 9)}$ (d) $F(s) = \dfrac{-2s + 8}{s^2 + 11s + 24}$

4–11   Find the time functions of the following transforms.

(a) $F(s) = \dfrac{9s - 18}{s(s + 2)(s^2 + 16)}$ (b) $F(s) = \dfrac{s^2 + 2s + 8}{(s + 1)(s^2 + 8s + 41)}$

(c) $F(s) = \dfrac{2s^2 - 4s + 1}{s^3 + 9s}$ (d) $F(s) = \dfrac{2s^3 - 1}{s^3 + 2s^2 + 5s}$

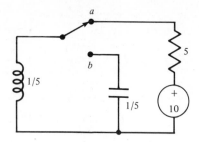

FIGURE 4–11.   Network for Problem 4–14.

4–12   Given the transform

$$F(s) = \frac{5s - 2}{s(s^2 + 4s + 16)}$$

show that there are complex poles and find the complete time function using the method of complex amplitudes and then using the method of matching.

4–13   Find the complete time function, including amplitudes, for the Laplace transform given in Problem 4–9.

4–14   In the network of Fig. 4–11 the switch moves from position *a* to position *b* at $t = 0$. Find the time function of the current using Laplace transforms. Describe the result.

4–15   Repeat Problem 4–14 but assume that the inductor has a resistance of 1 Ω. How is the resulting current function changed?

4–16   Suppose in Problem 4–14 that the switch moves back to position *a* after 1.5 s in position *b*. Find the time voltage across the inductor after the switch has moved back to position *a*. (*Hint:* Use the results of Problem 4–14 to determine the initial conditions for this new problem with the time origin defined to be the moment the switch moves back to position *a*.)

4–17   The switch in the network of Fig. 4–12 closes at $t = 0$. Find and plot the output voltage. Describe this voltage.

4–18   In the network of Fig. 4–13 the switch closes at $t = 0$. Find the output voltage if $\omega = 3$ rad/s. Plot the result.

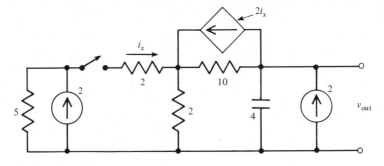

FIGURE 4–12.   Network for Problem 4–17.

4–19   For a general ω the steady-state solution of the network in Fig. 4–13 will contain such terms as

$$A \sin (\omega t) + B \cos (\omega t)$$

The amplitude of the oscillation is given by the expression $(A^2 + B^2)^{1/2}$. Find a general expression for this amplitude and plot the result for ω varying from 0 to 4 rad/s. Describe the response of the network.

4–20   Given the Laplace transform

$$F(s) = \frac{-3s^3 + 22s^2 + 201s + 340}{s^4 + 12s^3 + 57s^2 + 126s + 104}$$

a computer program was used to find the poles of the denominator as $p_1 = -2$, $p_2 = -4$, $p_3 = -3 - 4j$, and $p_4 = -3 + 4j$. Find the time function.

COMPUTER PROBLEMS

4–21   Find the poles and zeros of the transform

$$F(s) = \frac{2s^4 + 2s^3 - 2s^2 + 8s - 48}{s^5 + 6s^4 + 97s^3 + 322s^2 + 2206s + 3816}$$

Specify the form of the time dependence.

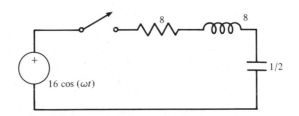

FIGURE 4–13.   Network for Problems 4–18 and 4–19.

4-22   Given the transform

$$V(s) = \frac{52s^3 - 8s + 1}{s^4 + (6 - 2K)s^3 + (37 - 8K + K^2)s^2 + (58 - 8K + 2K^2)s}$$

find a value for K that will give a steady-state oscillation in the time function. Specify the frequency and the form of the overall time function at this value of K. (*Hint:* Find the poles as different values of K are assigned.)

CHAPTER 5

General Solution
of Linear Networks

Objectives

After reading this chapter and working the problems at the end of the chapter, you will be able to:

1. Transform a network to s-space, including initial conditions, elements, and sources.
2. Find the matrix equation in s-space of a network using generalized mesh or nodal analysis.
3. Solve the matrix equation for the Laplace transform of the unknown or unknowns desired.
4. Invert the transform to find the time functions of the unknown or unknowns desired.
5. Use Laplace transforms to solve for the response of a network and make interpretations of network behavior based on the solution.

INTRODUCTION

The global objective of this chapter is to bring together all the network analysis procedures learned in previous chapters. The result is a *generalized procedure* which can be used to solve any linear network. If a network contains nonlinear elements, they must be replaced by linear models which are valid over the variable range for which the problem is to be solved. The method to be developed uses a matrix representation of the equations that will solve the network. The great simplification provided by this method is that the matrix can be constructed, almost from inspection of the network, even though capacitors, inductors, and transformers may be present. This is possible because the entire network is transformed into the *s*-space before any attempt is made to determine the equations. The Laplace equations are written directly from the network. Generalized network analysis involves the following operations,

1. Preparation. The network must be prepared for matrix equation construction. This is accomplished by determination of the initial conditions, modification of sources, and expression of elements in *s*-space.

2. Matrix Construction. A set of rules will be learned specifying how the matrix equation can be constructed from the network using mesh or nodal analysis. Adjustments of the matrix may be necessary to account for controlled sources.

3. Solution. Various techniques can be used to find solutions for particular unknowns in the network. Most common of these is Cramer's rule and determinant evaluation since the equations are in a matrix format. This step may or may not require inversion of the transform to find the time dependence.

4. Interpretation. A most important conclusion is that relating to the interpretations which can be given to the solutions obtained in network analysis.

PREPARATION

The first step in solving a network problem by the generalized mesh or nodal analysis method is to transform the *t*-space network into an *s*-space network. Such a transformation is carried out by three steps or operations:

1. Determine the initial conditions on capacitors and inductors and express as separate sources.

2. Replace all network components by a transformed representation of their impedance or admittance.
3. Replace all independent and controlled network sources by their Laplace transforms.

5-2.1 Initial Conditions

Chapters 3 and 4 showed that the initial conditions in networks are the voltages across capacitors and currents through inductors or transformers in the initial state of the network. These quantities are continuous across the time origin of the problem. They represent stored energy that will affect the later response of the network. In generalized mesh and nodal analysis these initial conditions will be treated as analytically equivalent sources, separated from the components themselves.

Equivalence is defined by specification that an ideal measurement made across the terminals of the component with initial condition is matched by the source and component model. An ideal measurement is one by which a measurement of some parameter is made independent of any other parameter: for example, the internal capacity is measured, regardless of any inductance, resistance, voltage source, or current source which may be present.

Capacitors. A capacitor has an initial condition consisting of the voltage across its terminals at the time origin of the problem. Figure 5–1a shows a capacitor with an initial charge, $v(0)$. The electrical characteristics of this component can be summarized as follows:

1. A measurement of open-circuit voltage will give a value of $v(0)$, since open-circuit means that no current is drawn and therefore the capacitor will not lose its charge.
2. A measurement of resistance will give infinity and inductance will give zero and a measurement of capacity will give C.

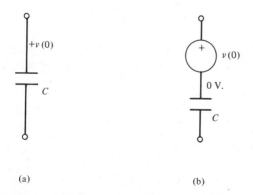

(a)   (b)

FIGURE 5–1. Representation of a charged capacitor by an uncharged capacitor and a voltage source.

3. If a resistance of magnitude R is placed across the terminals of the capacitor, you can see that the capacitor will discharge with a current

$$i(t) = \frac{v(0)}{R} e^{-t/RC}$$

and the voltage will be

$$v(t) = v(0)e^{-t/RC}$$

4. The total energy that can be drawn from the capacitor will be given by $\frac{1}{2}Cv^2(0)$.

Now consider the case of an *uncharged* capacitor in series with a voltage source of value $v(0)$ as shown in Fig. 1–5b. What electrical characteristics will such a system exhibit?

1. A measurement of open-circuit voltage will give a value of $v(0)$ since there is no voltage across the capacitor.
2. A measurement of resistance will give infinity and inductance will give zero since the sources has a zero resistance and inductance. A measurement of capacity will give C.
3. If a resistance of magnitude R is placed across the terminals of the system, a current will flow as the capacitor is charged through the resistor. You can show that the value of this current is

$$i(t) = \frac{v(0)}{R} e^{-t/RC}$$

and the capacitor plus source voltage is

$$v(t) = v(0)e^{-t/RC}$$

4. The total energy that can be drawn from the capacitor and source can be found by integrating the product of $i(t)$ and $v(t)$ from zero to infinity,

$$W = \int_0^\infty i(\tau)v(\tau)\, d\tau = \frac{v^2(0)}{R} \int_0^\infty e^{-\tau/RC}e^{-\tau/RC}\, d\tau$$

$$= \frac{1}{2}Cv^2(0)$$

Note that all four characteristics of the situations shown in Fig. 5–1a and b are identical. It is clear that the configuration shown in Fig. 5–1b can be used as an analytically equivalent model for a charged capacitor because there is no measurement that can distinguish between the two circuits. Thus, when using mesh analysis, all charged capacitors will be replaced by uncharged capacitors in series with a voltage source whose value and polarity matches that of the capacitor charge.

If nodal analysis is to be used for construction of the matrix equation, sources should be expressed as current sources. This is no problem if the voltage sources are shunted by resistors since the simple source transformation rule shows how the conversion is to be made. The model of a charged capacitor presents a problem, however, since there will be no shunted resistance. In this case the model is provided by the combination shown in Fig. 5–2. Note that the charge on the capacitor has been replaced by an *impulse* current source of magnitude given by $Cv(0)$:

$$i(t) = Cv(0)\delta(t) \tag{5-1}$$

To show that this model is equivalent, let us consider the four measurements above:

1. If a measurement of open-circuit voltage is made, all the current of the source will flow into the capacitor at $t = 0$. This will produce a voltage on the capacitor given by

$$v(t) = \frac{1}{C} \int_0^\infty i(\tau)\, d\tau = \frac{1}{C} \int_0^\infty Cv(0)\delta(\tau)\, d\tau$$

or

$$v(t) = v(0)$$

2. A measurement of resistance gives infinity and inductance will give zero since the current source has no inductance and is an open circuit. A measurement of capacity will give C.

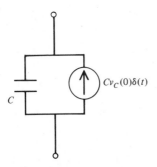

FIGURE 5–2. Representation of a charged capacitor by an uncharged capacitor and an impulse current source.

3. If a resistance of R ohms is placed across the circuit, a current will flow instantaneously into the capacitor charging it as given above, and then the capacitor will discharge as

$$i(t) = \frac{v(0)}{R} e^{-t/RC}$$

and voltage of

$$v(t) = v(0)e^{-t/RC}$$

4. The total energy drawn will be $\frac{1}{2}Cv^2(0)$, since this energy flows into the capacitor at $t = 0$ and is then drawn off by whatever load is connected.

So all four measurements will again result in the same values. This means that when nodal analysis is to be used a charged capacitor is replaced by an uncharged capacitor shunted by an impulse current source given by Eq. (5-1).

Inductor. The initial condition on an inductor (and for the transformer) will be an initial current. Figure 5–3a shows an inductor with an initial current $i(0)$. The electrical characteristics of this system are:

1. A measurement of current through a short placed across the terminals of the inductor will give an instantaneous value of $i(0)$.
2. A measurement of resistance will give zero while the capacity will be zero. The inductance will be L.
3. If a resistor R is placed across the terminals, there will be a current given by

$$i(t) = i(0)e^{-Rt/L}$$

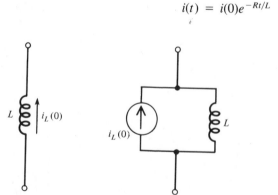

(a)   (b)

FIGURE 5–3. Representation of an inductor with an initial current by an inductor with no current shunted by a current source.

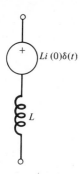

$Li(0)\delta(t)$

L

FIGURE 5—4. **Representation of an inductor with an initial current by an inductor with no current in series with an impulse voltage source.**

while the voltage across the terminals will be

$$v(t) = -Ri(0)e^{-Rt/L}$$

4. The total energy that can be drawn from the system is given by $\frac{1}{2}Li^2(0)$.

Now, consider the model shown in Fig. 5–3b, where the current has been presented as a current source shunting the inductor, which now has no initial current. As was done for the capacitor, you can show that measurements of the four quantities given above will turn out to be exactly the same for this model. Thus when doing nodal analysis an inductor with an initial current will be modeled as an inductor with no current shunted by a source with magnitude and direction to match the initial condition.

In the case of mesh analysis, all sources should be represented as voltage sources and the model of Fig. 5–3b would not be satisfactory. In this case you can be shown that the model presented in Fig. 5–4 will also have all the electrical characteristics given above. Thus when using mesh analysis the inductor will be modeled by an inductor with no initial condition in series with an impulse voltage source:

$$v(t) = Li(0)\delta(t) \tag{5-2}$$

EXAMPLE 5-1
Given the network of Fig. 5–5, find the initial conditions and show how they would be expressed as sources for mesh and nodal analysis.

SOLUTION
The initial conditions are established for the network in the configuration *before* the switch is closed. In this case there are actually two separate networks, as shown in Fig. 5–6, and each will establish initial conditions. For the 4-H inductor it is clear that there is a counterclockwise current of $i_{4H}(0) = 2$ A due

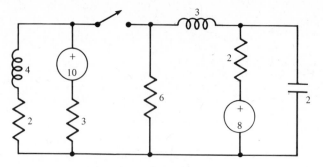

FIGURE 5–5. Network for Example 5–1.

to the 10-V source and total of 5 Ω. For the network on the right there is a counterclockwise current through the 3-H inductor of $i_{3H}(0) = 1$ A, due to the 8-V source and 8-Ω resistance in this loop. The capacitor charge is given by the 8-V source and the voltage drop acorss the 2-Ω resistor:

$$v_{2F}(0) = 8 + 2(-1) = 6 \text{ V}$$

Figure 5–7 shows the network with these initial conditions separated from the components by models of current and voltage sources. Using the definitions given above, this network can be configured as Fig. 5–8 for mesh analysis, and everything as voltage sources or as Fig. 5–9 for nodal analysis and everything as current sources. Note the delta functions for the impulse initial conditions. ∎

EXAMPLE 5-2

The network in Fig. 5–10 uses a transformer. Find the initial conditions and express as voltage sources as preparation for using mesh analysis.

SOLUTION

In Fig. 5–11 the network has been drawn in the initial state, prior to the switch being closed. The controlled source is gone since there is no current

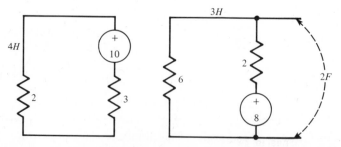

FIGURE 5–6. Network for finding the initial conditions in the network of Fig. 5–5.

GENERAL SOLUTION OF LINEAR NETWORKS

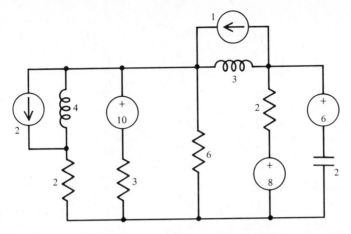

FIGURE 5–7. Network of Fig. 5–5 with initial conditions expressed as sources.

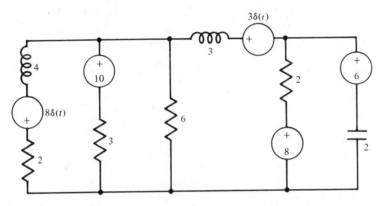

FIGURE 5–8. All sources in the network of Fig. 5–7 have been expressed as voltages sources for use of mesh analysis.

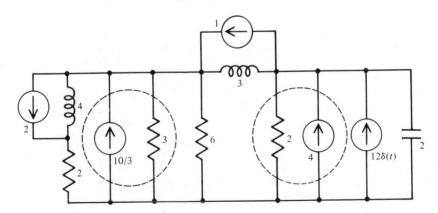

FIGURE 5–9. All sources in the network of Fig. 5–7 have been expressed as current sources for use of nodal analysis.

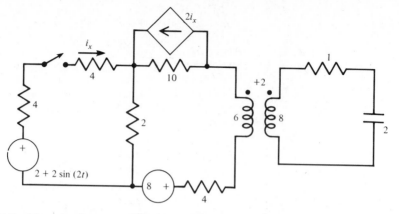

FIGURE 5–10. Network for Example 5–2.

through the 4-Ω resistor and therefore no current generated by the source. It is clear from this network that there is a counterclockwise current of $8/(10 + 2 + 4) = 0.5$ A. Note that the T model of the transformer is being used. Since there is no dc voltage drop across an inductor, there is no voltage on the capacitor and no current through the 6-H part of the T. So the initial conditions are

$$i_{4H}(0) = -0.5 \text{ A}$$

$$i_{2H}(0) = -0.5 \text{ A}$$

$$i_{6H}(0) = 0 \text{ A}$$

$$v_{2F}(0) = 0 \text{ V}$$

The current sources are shunted across each inductor, as shown in Fig. 5–12. They are then transformed into series impulse voltage sources to give the final network shown in Fig. 5–13. ∎

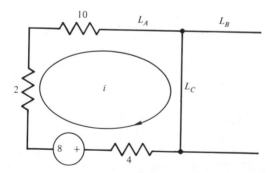

FIGURE 5–11. Initial state of the network of Fig. 5–10 for finding the initial conditions.

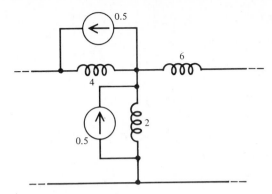

FIGURE 5–12. Transformer initial conditions expressed as current sources.

5-2.2 Transformed Components

The next step in the preparation of the network for generalized mesh or nodal analysis is to transform the components from *t*-space to *s*-space. This is accomplished by defining the *transformed impedance Z(s)* and *transformed admittance Y(s)* of network components.

Suppose that some component has a voltage drop of $v(t)$ and a current of $i(t)$. Then the relationship between these quantities may involve proportions, derivatives, or integrals, depending on the type of component. This can be seen from Table 3–1. However, it was shown that all of these operations reduce to algebra when Laplace transforms are taken. Thus if only the transforms are considered, $V(s)$ and $I(s)$, it makes sense to consider also the relationship between these quantities. Such a relationship is defined by transforming the time relation for the components. The transformed impedance is therefore defined as

$$Z(s) = \frac{V(s)}{I(s)} \tag{5-3}$$

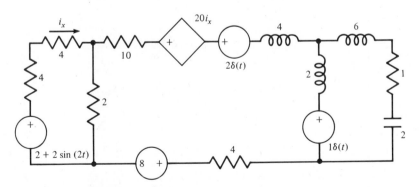

FIGURE 5–13. Final form of the network in Fig. 5–10, set up for use of mesh analysis with only voltage sources.

and the transformed admittance is given by

$$Y(s) = \frac{1}{Z(s)} = \frac{I(s)}{V(s)} \tag{5-4}$$

It remains to define the ratio of $V(s)$ and $I(s)$ for each component.

Resistor.   For the resistor, Ohm's law gives the required relation in time:

$$v_R(t) = Ri_R(t)$$

If the Laplace transform is taken of this equation, the result is

$$V_R(s) = RI_R(s)$$

from which it is clear that the transformed impedance for a resistor is given by

$$Z_R(s) = R \tag{5-5}$$

and of course the transformed admittance is

$$Y_R(s) = G = \frac{1}{R} \tag{5-6}$$

Capacitor.   For the capacitor it will be assumed that any initial condition has been extracted as a source, leaving the capacitor uncharged. Then, from Table 3–1, the capacitor voltage can be written

$$v_C(t) = \frac{1}{C} \int_0^t i_C(\tau)\, d\tau$$

Taking the Laplace transform of this equation gives

$$V_C(s) = \frac{I_C(s)}{sC}$$

From this equation it is clear that the transformed impedance of a capacitor is

$$Z_C(s) = \frac{1}{sC} \tag{5-7}$$

and of course the transformed admittance is

$$Y_C(s) = sC \tag{5-8}$$

Inductor. The inductor follows a relationship between voltage and current given by

$$v_L(t) = L \frac{di_L}{dt}$$

Under the assumption that any initial condition on the inductor has been expressed as an external source, the Laplace transform of this expression yields

$$V_L(s) = sLI_L(s)$$

From this expression it follows immediately that the transformed impedance of an inductor is

$$Z_L(s) = sL \tag{5-9}$$

and the transformed admittance is given by

$$Y_L(s) = \frac{1}{sL} \tag{5-10}$$

Transformer. The appropriate s-space relations for a transformer can be derived by using a T model, as given in Chapter 1, which expresses the element in terms of inductors. In this case the relations of Eq. (5-9) or (5-10) can be used to deduce the transformed impedance or admittance. If the direct relations for a transformer as equations for primary and secondary current and voltage are used, generalized mesh and nodal analysis cannot be used. In these cases it will be necessary to use the KVL or KCL approaches to solving the network as described in Chapter 4.

5-2.3 Transformed Sources

The last step in preparation of the network is to replace all real and controlled sources by their Laplace transforms. This includes those sources which resulted from extraction of initial conditions from capacitors, inductors, and transformers.

Real Sources. The Laplace transform of real sources is provided by looking up the transform of whatever time function describes the source. Table 4–1 is used to deduce the appropriate transform. Source polarity or direction is preserved in such a transformation.

Controlled Sources. Controlled sources in t-space will still be controlled sources in s-space. After transformation, however, the dependency will be in terms of the Laplace transform of some current or voltage in the network. For example, suppose that an ICVS has the given dependence

$$v(t) = 10i_x(t)$$

In the transformed network the ICVS relationship would take on the form

$$V(s) = 10I_x(s)$$

so that if $i_x(t)$ is some current in the network, $I_x(s)$ is the Laplace transform of that current.

5-2.4   Transformed Network

By combining the transformations presented in the previous sections, it is now possible to effectively construct the *Laplace transform* of the network itself. Instead of deriving the time equations from the network and then taking the Laplace transform, the transform is taken first; then the Laplace equations can be constructed directly from the network. The following examples show the forms that a transformed network may take.

EXAMPLE 5-3

Construct the transformed network for the *t*-space network of Example 5-1 as described by Figs. 5–8 and 5–9.

SOLUTION

In Example 5-1 the initial conditions of the network given in Fig. 5–5 were calculated. These initial conditions were expressed as sources and then everything was expressed as voltage sources in the representation of the network given in Fig. 5–8. Laplace transformation of the network is accomplished by replacing every element by its transform. In this case transformed impedance will be used in anticipation of solving the problem with mesh analysis. Thus the 4-H inductor becomes a $4s$ inductor, the 2-F capacitor becomes a $1/2s$ capacitor, and resistors stay the same. The impulse sources transform to a constant. Then the $8\delta(t)$ sources becomes simply an 8 with the same polarity. The 10-V source becomes $10/s$, and so on. The final result is shown in Fig. 5–14.

For the case of Fig. 5–9, the problem has been set up for generalized nodal analysis since current sources have been used. In this case the components will be replaced by transformed admittance. The 4-H inductor will become a $1/4s$ inductor, the 2-F capacitor becomes $2s$, and the resistors are replaced by their inverse values. Sources are transformed according to their stated time dependence. The resulting network is shown in Fig. 5–15. ∎

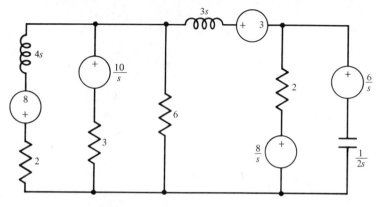

FIGURE 5—14. Transformed network of Fig. 5—8, using transformed impedences as required for mesh analysis.

EXAMPLE 5-4

Find the transformed network for the *t*-space network given in Fig. 5–16. Use a nodal representation.

SOLUTION

First the initial conditions must be determined and expressed as sources. This is done using the network in the initial state (i.e., with the switch open). The dc network is shown in Fig. 5–17. Note that the 4-H inductor, a dc short circuit, has effectively eliminated the shunting 4-Ω resistor and the 10-V source and 5-Ω resistor from the dc network. *However,* the 10-V source and 5-Ω resistor do provide a 2-A current clockwise through this inductor. This means that the total current through this inductor will be the 2 A plus whatever i_2 turns out to be. Clearly, the 2-F capacitor will be uncharged, $v_{2F}(0) = 0$. The network of

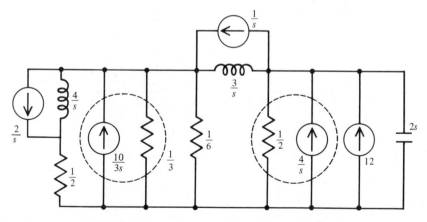

FIGURE 5—15. Transformed network of Fig. 5—9, using transformed admittances as required for nodal analysis.

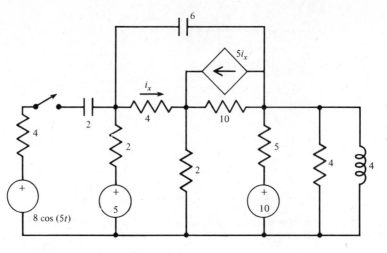

FIGURE 5–16. Network for Example 5-4.

Fig. 5–17 must now be solved for i_1 and i_2 to deduce the remaining initial conditions.

$$\begin{pmatrix} 8 & -2 \\ -2 & 12 \end{pmatrix}\begin{pmatrix} i_1 \\ i_2 \end{pmatrix} = \begin{pmatrix} 5 \\ -50i_x \end{pmatrix}$$

but $i_x = i_1$, so

$$\begin{pmatrix} 8 & -2 \\ 48 & 12 \end{pmatrix}\begin{pmatrix} i_1 \\ i_2 \end{pmatrix} = \begin{pmatrix} 5 \\ 0 \end{pmatrix}$$

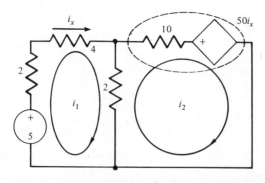

FIGURE 5–17. Network of Fig. 5–16 in the initial state.

Solving, we have

$$i_1 = \frac{\begin{vmatrix} 5 & -2 \\ 0 & 12 \end{vmatrix}}{\begin{vmatrix} 8 & -2 \\ 48 & 12 \end{vmatrix}} = \frac{60}{192} = 0.3125 \text{ A}$$

$$i_2 = \frac{\begin{vmatrix} 8 & 5 \\ 48 & 0 \end{vmatrix}}{192} = \frac{-240}{192} = -1.25 \text{ A}$$

From these two currents the initial conditions can be established,

$$i_{4H}(0) = 2 + i_2 = 2 - 1.25 = 0.75 \text{ A}$$

$$v_{6F}(0) = 4i_1 + 10i_2 + 50i_x = 10i_2 + 54i_1$$

$$= 10(-1.25) + 54(0.3125) = 3.125 \text{ V}$$

Figure 5–18 shows the network with these initial conditions inserted as sources. Now, to use the nodal representation, it will be necessary to convert the voltage sources to current sources. This will result in an impulse function in the case of the capacitor. Then all components are replaced by their transformed admittance and sources by their Laplace transforms. The resulting network is shown in Fig. 5–19. Notice that the controlled source dependency is retained. ■

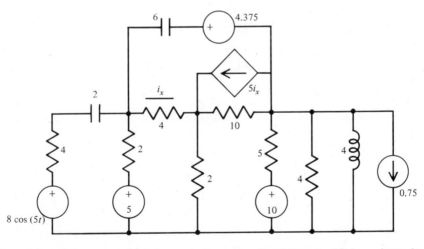

FIGURE 5–18. The initial conditions of the network of Fig. 5–16 have been inserted as sources.

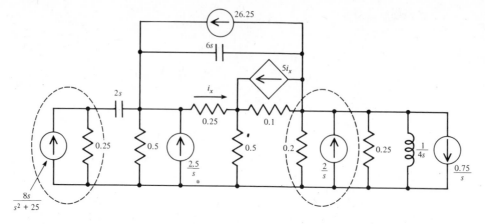

FIGURE 5–19. Transformed network of Fig. 5–16 ready for nodal analysis.

5-3

GENERALIZED NETWORK MATRIX EQUATIONS

Now that the network has been transformed into s-space it is possible to construct the matrix equation, which is the set of simultaneous equations necessary to solve for the network variables. Once the matrix equation has been constructed, a solution for the transform of network variables can be derived using the traditional methods such as Cramer's rule. Matrix construction follows a set of rules similar to those for the resistive mesh and nodal techniques developed in Chapter 2.

5-3.1 Generalized Mesh Analysis

To construct a matrix equation for the transformed network using mesh analysis means that the network variables solved for will be currents. The rules for matrix construction are as follows:

1. Identify the elementary loops. Assign to each loop an unknown clockwise current.
2. The matrix equation will have the form

$$
\begin{pmatrix}
z_{11} & z_{12} & \cdot & \cdot & \cdot & z_{1n} \\
z_{21} & z_{22} & \cdot & \cdot & \cdot & z_{2n} \\
z_{31} & \cdot & \cdot & \cdot & \cdot & \cdot \\
\cdot & \cdot & \cdot & \cdot & \cdot & \cdot \\
\cdot & \cdot & \cdot & \cdot & \cdot & \cdot \\
\cdot & \cdot & \cdot & \cdot & \cdot & \cdot \\
z_{n1} & z_{n2} & \cdot & \cdot & \cdot & z_{nn}
\end{pmatrix}
\begin{pmatrix}
I_1 \\
I_2 \\
\cdot \\
\cdot \\
\cdot \\
\cdot \\
I_n
\end{pmatrix}
=
\begin{pmatrix}
V_1 \\
V_2 \\
\cdot \\
\cdot \\
\cdot \\
\cdot \\
V_n
\end{pmatrix}
\qquad (5\text{-}11)
$$

3. $I_1, I_2, \ldots, I_n$ = unknown Laplace loop currents.

4. z_{ii} = diagonal elements = sum of the transformed impedances in loop i.

5. $z_{ij} = z_{ji}$ = off-diagonal elements = negative of transformed impedances shared by loops i and j. Note that the matrix is symmetric.

6. $V_1, V_2, \ldots, V_n$ = forcing functions = sum of transformed sources in each loop. If the transformed current leaves the positive terminal of the source, it is added to the term, and if the transformed loop current enters the positive terminal of the source, it is subtracted from the term.

These rules show how the matrix equation is constructed from a transformed network. The following examples illustrate such construction.

EXAMPLE 5-5

The network shown in Fig. 5–5 was shown to result in a transformed network, suitable for mesh analysis, as given in Fig. 5–14. Find the matrix equation for this network.

SOLUTION

In Fig. 5–20 the transformed network has been redrawn with the elementary loops identified and a clockwise current assigned to each. Using the rules given above, the resulting matrix equation is

$$
\begin{pmatrix}
5 + 4s & -3 & 0 & 0 \\
-3 & 9 & -6 & 0 \\
0 & -6 & 8 + 3s & -2 \\
0 & 0 & -2 & 2 + \dfrac{1}{2s}
\end{pmatrix}
\begin{pmatrix}
I_1 \\
I_2 \\
I_3 \\
I_4
\end{pmatrix}
=
\begin{pmatrix}
-8 - \dfrac{10}{s} \\
\dfrac{10}{s} \\
-3 - \dfrac{8}{s} \\
\dfrac{2}{s}
\end{pmatrix}
\quad \blacksquare
$$

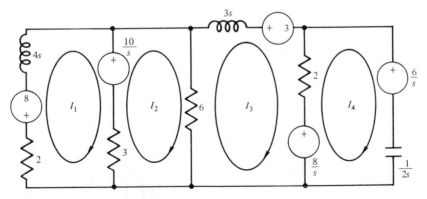

FIGURE 5–20. Transformed network of Example 5-5, prepared for generalized mesh analysis, with loop currents defined.

5-3.2 Generalized Nodal Analysis

When a network is to be analyzed using generalized nodal analysis, all the sources must be expressed as current sources. The network is prepared with all initial conditions expressed as current sources and the components described by their transformed admittance. Then the matrix equation for the node voltages is constructed as follows.

1. Identify all of the nodes, select one as the reference, and assign to the rest unknown s-space node voltages.
2. The matrix equation will have the form

$$
\begin{pmatrix}
y_{11} & y_{12} & \cdot & \cdot & \cdot & y_{1n} \\
y_{21} & y_{22} & \cdot & \cdot & \cdot & y_{2n} \\
y_{31} & \cdot & \cdot & \cdot & \cdot & \cdot \\
\cdot & \cdot & \cdot & \cdot & \cdot & \cdot \\
\cdot & \cdot & \cdot & \cdot & \cdot & \cdot \\
y_{n1} & y_{n2} & \cdot & \cdot & \cdot & y_{nn}
\end{pmatrix}
\begin{pmatrix}
V_1 \\
V_2 \\
\cdot \\
\cdot \\
\cdot \\
V_n
\end{pmatrix}
=
\begin{pmatrix}
I_1 \\
I_2 \\
\cdot \\
\cdot \\
\cdot \\
I_n
\end{pmatrix}
\tag{5-12}
$$

3. $V_1, V_2, \ldots, V_n$ = unknown node voltage transforms.
4. y_{ii} = diagonal elements = sum of transformed admittances connected to node i.
5. y_{ij} = off-diagonal elements = negative of transformed admittances joining nodes i and j. The matrix is symmetric, $y_{ij} = y_{ji}$.
6. $I_1, I_2, \ldots, I_n$ = forcing functions = sum of current sources connected to the nodes. Current sources directed into the node are added to the term and current sources directed out of the node are subtracted from the term.

The following example illustrates application of these rules to the construction of the matrix equations.

EXAMPLE 5-6

The network of Fig. 5–16 was shown to have a transformed network, suitable for nodal analysis, as given in Fig. 5–19. Construct the matrix equation for this network.

SOLUTION

In Fig. 5–21 the nodes have been identified and one selected as a reference. The other nodes have been assigned unknown s-space voltages, V_1 to V_4. The matrix equation constructed by the rules above is

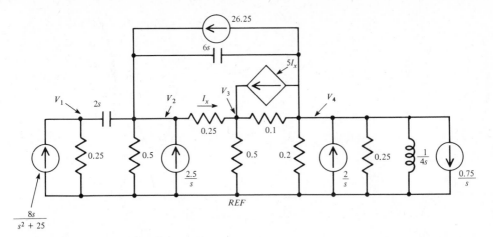

FIGURE 5–21. Transformed network of Example 5-6, prepared for generalized nodal analysis, with node voltages defined.

$$
\begin{pmatrix}
0.25 + 2s & -2s & 0 & 0 \\
-2s & 0.75 + 8s & -0.25 & -6s \\
0 & -0.25 & 0.85 & -0.1 \\
0 & -6s & -0.1 & 0.55 + \dfrac{1}{4s} + 6s
\end{pmatrix}
\begin{pmatrix}
V_1 \\
V_2 \\
V_3 \\
V_4
\end{pmatrix}
$$

$$
=
\begin{pmatrix}
\dfrac{8s}{s^2 + 25} \\[2mm]
\dfrac{2.5}{s} + 26.25 \\[2mm]
5I_x \\[2mm]
-5I_x - 26.25 + \dfrac{1.25}{2}
\end{pmatrix}
$$

From an examination of the network as given in Fig. 5–21 you can see that $I_x = (V_2 - V_3)/4$ or $5I_x = 1.25V_2 - 1.25V_3$. When this is substituted into the matrix equation and the terms are rearranged the resulting, final matrix equation is

$$\begin{pmatrix} 0.25 + 2s & -2s & 0 & 0 \\ -2s & 0.75 + 8s & -0.25 & -6s \\ 0 & 1.5 & 2.1 & -0.1 \\ 0 & 1.25 - 6s & -1.35 & 0.55 + \dfrac{1}{4} + 6s \end{pmatrix} \begin{pmatrix} V_1 \\ V_2 \\ V_3 \\ V_4 \end{pmatrix} = \begin{pmatrix} \dfrac{8s}{s^2 + 25} \\ \dfrac{2.5}{s} + 26.25 \\ 0 \\ \dfrac{1.25}{s} - 26.25 \end{pmatrix} \blacksquare$$

5-4

SOLUTION OF THE MATRIX EQUATION

When the matrix equation for some network has been constructed, the problem has been reduced to the mathematics of solving the equation for the unknown or unknowns desired. This will involve the same steps outlined in Chapter 4, which can be summarized as follows:

1. Obtain the Laplace transform for the unknown desired. This will usually involve the use of Cramer's rule to set up the solution as the quotient of two determinants. When these determinants are multiplied out, the result will be the ratio of two functions of s.
2. Place the resulting transform in good form, as the ratio of two polynomials in positive powers of s.
3. Verify that the resulting expression satisfies the highest power test. If not, divide until the remainder satisfies this test.
4. Obtain the roots of the polynomial in the denominator, which are the poles of the transform. Write the denominator in factored form using these poles.
5. Perform partial fraction expansion to express the transform as the sum and difference of simple fractions in the factors of the denominator.
6. Using the transform tables, invert each term in the expansion to find the final time function.

These same steps were used in Chapter 4 to find the time function of network problems using KVL and KCL to derive the equations. In general, for problems of even modest complexity, the use of a computer is essential for the efficient evaluation of the transform. Programs are available to find the roots of polynomials and to find the amplitudes in the partial fraction expansion. Often the network analysis is used to analyze the behavior of a network with different input signals. In this type of problem it is often not necessary to obtain a complete solution to the problem to provide the required answers. The following example illustrates such a problem and its solution.

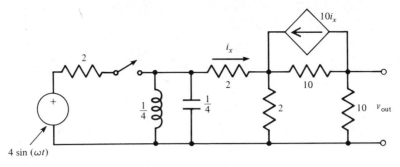

FIGURE 5–22. Network for Example 5-7.

EXAMPLE 5-7

The network in Fig. 5–22 is a model of a tuned amplifier. The switch applies an input signal of some frequency ω to the circuit. The $\frac{1}{4}$-H inductor and $\frac{1}{4}$-F capacitor form what is called a tuned circuit at the input of a simple amplifier composed of an ICIS. The problem is to determine the output voltage as a function of input signal frequency.

SOLUTION

This problem can be solved using generalized network analysis, but with the input signal frequency left arbitrary, represented by ω. Following the normal procedures it is first noted that there are no initial conditions since in the initial state no real power sources are connected to the capacitor or inductor. The next step is to transform the network to s-space. Since a parallel arrangement of elements is indicated, nodal analysis will be used. In Fig. 5–23 the network has been transformed to s-space. Note that the input voltage source has been transformed into a current source. There are three nodes and the reference. The

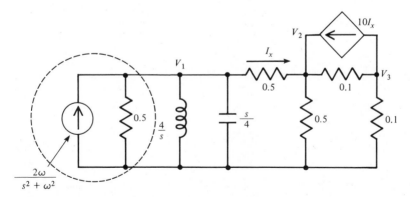

FIGURE 5–23. Network of Fig. 5–22, transformed and prepared for generalized nodal analysis.

desired output voltage will be node 3 voltage, $V_3(s)$. By the nodal matrix construction rules the matrix equation becomes

$$\begin{pmatrix} 1 + \dfrac{4}{s} + \dfrac{s}{4} & -0.5 & 0 \\ -0.5 & 1.1 & -0.1 \\ 0 & -0.1 & 0.2 \end{pmatrix} \begin{pmatrix} V_1(s) \\ V_2(s) \\ V_3(s) \end{pmatrix} = \begin{pmatrix} \dfrac{2\omega}{s^2 + \omega^2} \\ 10I_x \\ -10I_x \end{pmatrix}$$

It is clear that $I_x = (V_1 - V_2)/2 = 0.5V_1 - 0.5V_2$. When this is inserted into the matrix equation the resulting relationship is

$$\begin{pmatrix} 1 + \dfrac{4}{s} + \dfrac{s}{4} & -0.5 & 0 \\ -5.5 & 6.1 & -0.1 \\ 5 & -5.1 & 0.2 \end{pmatrix} \begin{pmatrix} V_1(s) \\ V_2(s) \\ V_3(s) \end{pmatrix} = \begin{pmatrix} \dfrac{2\omega}{s^2 + \omega^2} \\ 0 \\ 0 \end{pmatrix}$$

Cramer's rule can now be used to solve for $V_3(s)$:

$$V_3(s) = \frac{\begin{vmatrix} 1 + \dfrac{4}{s} + \dfrac{s}{4} & -0.5 & \dfrac{2\omega}{s^2 + \omega^2} \\ -5.5 & 6.1 & 0 \\ 5 & -5.1 & 0 \end{vmatrix}}{\begin{vmatrix} 1 + \dfrac{4}{s} + \dfrac{s}{4} & -0.5 & 0 \\ -5.5 & 6.1 & -0.1 \\ 5 & -5.1 & 0.2 \end{vmatrix}}$$

When this expression is multiplied out and simplified, the following transform results:

$$V_3(s) = \frac{-27.61\omega s}{(s^2 + \omega^2)(s^2 + 2.31s + 16)}$$

The poles of the quadratic turn out to be given by poles $= -1.155 \pm 3.83j$, a complex pair. This means that the partial fraction expansion can be written as

$$V_3(s) = \frac{\omega A}{s^2 + \omega^2} + \frac{Bs}{s^2 + \omega^2} + \frac{3.83C}{(s + 1.155)^2 + 14.7}$$

$$+ \frac{(s + 1.155)D}{(s + 1.155)^2 + 14.7}$$

Thus A and B are sin (ωt) and cos (ωt) amplitudes and C and D are the $e^{-1.155t}$ sin $(3.83t)$ amplitude and $e^{-1.155t}$ cos $(3.83t)$ amplitude. Now it is only necessary to determine the A and B amplitudes as a function of ω. Using the method of matching, equations for these amplitudes can be derived. If the equation above is placed over a common denominator and the numerator equated to the numerator of the original transform, the following matrix equation for the amplitudes results:

$$
\begin{pmatrix}
0 & 1 & 0 & 1 \\
\omega & 2.31 & 3.83 & 1.155 \\
2.31\omega & 16 & 0 & \omega^2 \\
16\omega & 0 & 3.83\omega^2 & 1.155\omega^2
\end{pmatrix}
\begin{pmatrix} A \\ B \\ C \\ D \end{pmatrix}
=
\begin{pmatrix} 0 \\ 0 \\ -27.61\omega \\ 0 \end{pmatrix}
$$

This equation can be solved for A and B using Cramer's rule. The result will be the amplitudes of the input signal on the output, as a function of the input signal frequency. This means that, after the transient oscillation has died away, the output will be

$$
v_{out}(t) = A \sin (\omega t) + B \cos (\omega t)
$$

or, expressed as amplitude and phase,

$$
v_{out}(t) = (A^2 + B^2)^{1/2} \sin (\omega t + \phi)
$$

where

$$
\phi = \tan^{-1}\left(\frac{B}{A}\right)
$$

The solutions of A and B are found to be

$$
A = \frac{-63.8\omega^2}{\omega^4 - 26.7\omega^2 + 256}
$$

$$
B = \frac{27.61\omega(\omega^2 - 16)}{\omega^4 - 26.7\omega^2 + 256}
$$

Figure 5–24 shows graphs of A and B versus frequency. Notice the A amplitude peak at $\omega = 4$. This shows that the amplifier has a gain which is frequency dependent. Figure 5–25 shows a graph of the net signal amplitude and phase shift versus frequency. Notice that the amplitude is 12 at $\omega = 4$ for a gain of $12/4 = 3$ with a zero phase shift. ■

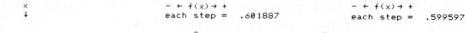

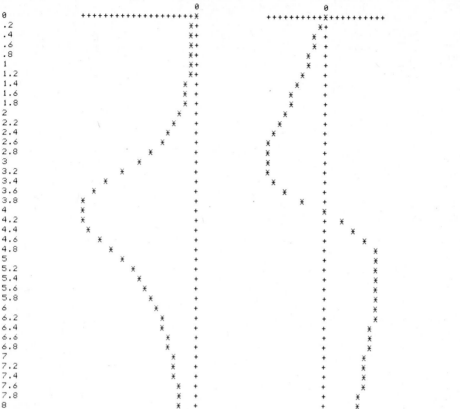

FIGURE 5–24.   Sine and cosine output voltage amplitude as a function of frequency for the network of Fig. 5–22.

This example illustrates how the generalized network analysis techniques can be used to look for particular types of solutions of network problems. The next section considers yet another application.

5·5

INTERPRETATIONS

The application of generalized mesh and nodal analysis can provide a complete solution to any linear network. The amount of calculation that may be required to obtain a final solution, consisting of the time-dependent functions of the network variables, can be substantial. It is very important to realize that much of the value of Laplace transforms, and generalized network analysis, lies in the conclusions that can be drawn from preliminary results, without the need to

GENERAL SOLUTION OF LINEAR NETWORKS

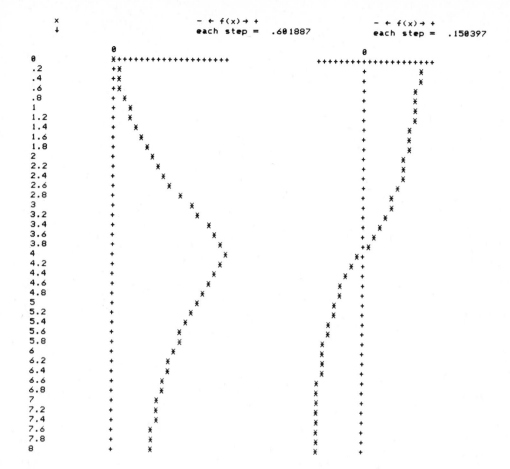

FIGURE 5—25. Net output voltage amplitude and phase versus frequency for the network of Fig. 5—22.

obtain a complete solution to the problem. In this section a number of ways of interpreting the results of network analysis will be discussed.

5-5.1 Time Dependence

In Chapter 4 it was shown that the functional form of the time dependence that some variable will have can be deduced as soon as the poles of the denominator are known. There are many problems for which this is the desired information. In these cases there is no need to evaluate the amplitudes of the terms. The type of information that can be obtained falls into several categories.

Transient Behavior. When the poles have been determined, the transient behavior of the network variables is determined by the presence of real poles and complex poles with a nonzero real part. It has been pointed out that the real poles will always be negative. This is because these poles will contribute exponential decay via terms like, e^{-at}. If the pole *was not* negative, the term would represent exponential growth instead of decay. This is physically not possible. If a voltage or current began to exhibit exponential growth, it would quickly reach a value causing some sort of circuit breakdown, or the linear model would become invalid to describe the network. From the negative poles the length of time that transients last can be estimated as approximately five times the largest time constant. Remember that the time constant is one over the decay constant, $\tau = 1/a$.

Steady-State Behavior.   The poles also allow a determination of the steady-state behavior of the network variables. This is the nature or form of the variables after all transients have died away. Of course, a pole at zero indicates the presence of a dc level. A complex pole with a zero real part represents a steady-state oscillation, at a frequency given by the complex part of the pole. Thus the frequency of oscillation is known when the poles have been identified.

When only the time response of a variable is desired, it is only necessary to calculate the transform to the point where it is expressed as the ratio of two polynomials. In that form the roots of the denominator can be determined and hence the poles.

5-5.2 Parametric Response

One of the most important applications of generalized network analysis is to study the effect of variation of some network parameter. Example 5-7 studied the effect of variation of input signal frequency on network response by finding how the gain and phase varied with frequency. This is an example of a study of parametric response. Often the critical question in a network is the value of gain associated with some controlled source. The magnitude of such a constant can have very profound effects on the network time response. The following two examples examine two different types of effect which controlled source gains can have.

EXAMPLE 5-8

In Fig. 5–26 a network is shown with a ICVS. The amplitude of this source is given by Ki_x, where i_x is identified in the network. The switch closes at $t = 0$. Describe the nature of the time response of the current through the inductor as a function of the controlled source gain.

SOLUTION

The solution of this problem will be found by finding the denominator of the loop current as a function of K and then deducing how the value of this constant affects the nature of the poles. First the network must be transformed to s-space and the matrix equation constructed. There are no initial conditions since no power is applied prior to the switch being closed. Figure 5–27 shows the trans-

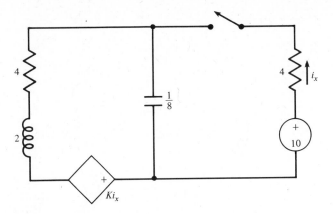

FIGURE 5—26.   Network for Example 5-8.

formed network. The matrix equation, after adjustment for the controlled source, has the form

$$
\begin{pmatrix} 2s + 4 + \dfrac{8}{s} & -\dfrac{8}{s} - K \\[4mm] -\dfrac{8}{s} & 4 + \dfrac{8}{s} \end{pmatrix} \begin{pmatrix} I_1 \\[4mm] I_2 \end{pmatrix} = \begin{pmatrix} 0 \\[4mm] -\dfrac{10}{s} \end{pmatrix}
$$

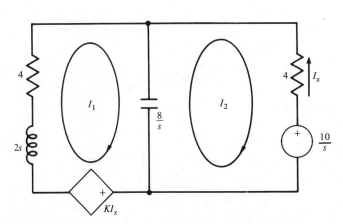

FIGURE 5—27.   Network of Fig. 5—26, transformed and prepared for generalized mesh analysis.

I_1 will give the current through the inductor:

$$
I_1 = \frac{\begin{vmatrix} 0 & -\dfrac{8}{s} - K \\[3mm] -\dfrac{10}{s} & 4 + \dfrac{8}{s} \end{vmatrix}}{\begin{vmatrix} 2s + 4 + \dfrac{8}{s} & -\dfrac{8}{s} - K \\[3mm] -\dfrac{8}{s} & 4 + \dfrac{8}{s} \end{vmatrix}}
$$

When this determinant ratio is multiplied out and expressed as the ratio of two polynomials, the result is

$$
I_1 = \frac{-10 - s(10K/8)}{s[s^2 + 4s + (8 - K)]}
$$

The roots of the quadratic can be easily shown to be given by

$$
\text{poles} = -2 \pm (K - 4)^{1/2}
$$

Therefore, the form of the time dependence is a function of the magnitude of K since the poles determine the time dependence. Let us examine the form of the functions that result for different ranges of values of K.

$K > 8$: For K larger than 8, one of the poles will be positive real, which is not a physical solution because it represents a growing exponential. The network model is not valid for values of K greater than 8.

$K = 8$: For K equal to 8, there is a repeated pole at zero and a pole at -4. The time function can be shown to be

$$
i(t) = -1.875 + 1.875t + 1.875e^{-4t}
$$

$4 < K < 8$: For K in this range, the poles are found to consist of the dc term and two decaying exponentials. For example, with $K = 4.25$ the solution is

$$
i(t) = -2.67 + 1.32e^{-2.5t} + 1.35e^{-1.5t}
$$

$K = 4$: The solution for K equal to 4 will have a repeated pole at zero, but the solution turns out to involve only one and a decaying exponential,

$$i(t) = -2.5 + 2.5e^{-2t} ,$$

$K < 4$: For K less than 4, including negative values, the solution will have a dc term and a decaying oscillation. The frequency of oscillation will increase with K magnitude but the decay time remains constant. For example, the solution for $K = 0$ is

$$i(t) = -1.25 + 1.25e^{-2t} \cos{(2t)} + 2.5e^{-2t} \sin{(2t)}$$
$$\text{while the solution for } K = -12 \text{ is}$$
$$i(t) = -0.5 + 0.5e^{-2t} \cos{(4t)} + 4e^{-2t} \sin{(4t)}$$

Figure 5-28 shows plots of the solutions for K equal to 8, 4, 0, and -12. Note the variation in form. ■

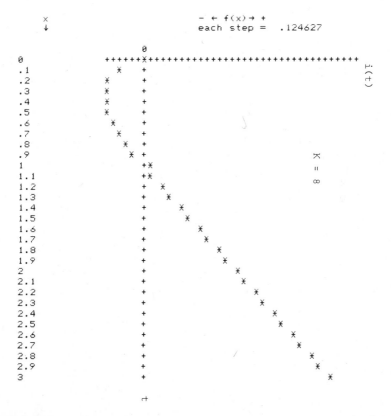

FIGURE 5–28 (a). **Effect of controlled source gain on the current in the network of Fig. 5–26: $K = 8$.**

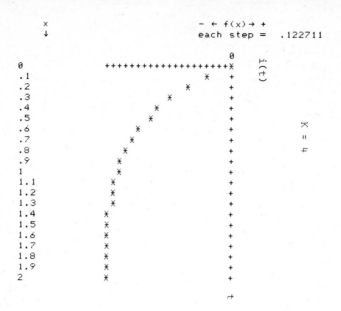

FIGURE 5—28 (b). *K* = 4.

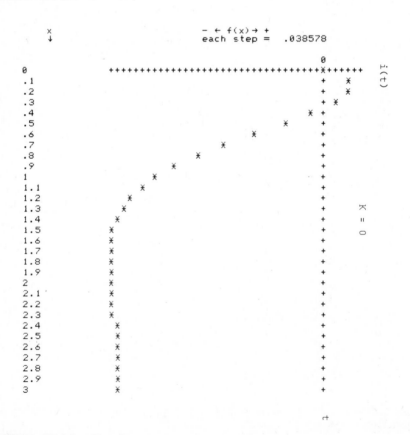

FIGURE 5—28 (c). *K* = 0.

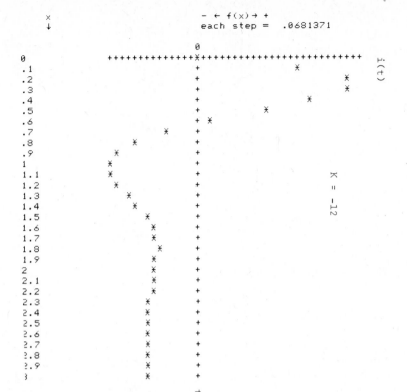

```
          x                    -  ← f(x) →  +
          ↓                    each step =   .0681371


                                    0
  0                     +++++++++++++++X+++++++++++++++++++++++++++++
  .1                                  +              *
  .2                                  +                     *         i(t)
  .3                                  +                     *
  .4                                  +                 *
  .5                                  +         *
  .6                                  + *
  .7                            *      +
  .8                        *          +
  .9                   *               +
  1                 *                  +
  1.1               *                  +                        K
  1.2                *                 +                        =
  1.3                 *                +                        -12
  1.4                 *                +
  1.5                   *              +
  1.6                    *             +
  1.7                    *             +
  1.8                      *           +
  1.9                    *             +
  2                      *             +
  2.1                    *             +
  2.2                    *             +
  2.3                   *              +
  2.4                   *              +
  2.5                   *              +
  2.6                   *              +
  2.7                   *              +
  2.8                   *              +
  2.9                   *              +
  3                    *               +

                                    ⊢
```

FIGURE 5–28 (d). $K = -12$.

EXAMPLE 5-9

Figure 5–29 shows a network slightly modified from Example 5-8. Again the form of the time dependence is to be studied as a function of the ICVS gain, K.

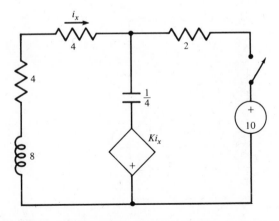

FIGURE 5–29. Network for Example 5-9.

SOLUTION

Again there are no initial conditions. The transformed network is shown in Fig. 5-30. The matrix equation is

$$\begin{pmatrix} 8 - K + 8s + \dfrac{4}{s} & -\dfrac{4}{s} \\[3mm] K - \dfrac{4}{s} & 2 + \dfrac{4}{s} \end{pmatrix} \begin{pmatrix} I_1 \\[3mm] I_2 \end{pmatrix} = \begin{pmatrix} 0 \\[3mm] -\dfrac{10}{s} \end{pmatrix}$$

When I_1 is solved for via Cramer's rule, the result is

$$I_1 = \frac{-2.5}{s[s^2 + (3 - 0.125K)s + 2.5]}$$

The poles of the denominator are the pole at zero and, from the quadratic,

$$\text{poles} = -\frac{(3 - 0.125K)}{2} \pm \frac{[(3 - 0.125K)^2 - 10]^{1/2}}{2}$$

Notice that in this case the gain changes the frequency, as in Example 5-8, but now also the damping coefficient (i.e., the real part of the pole). A very interesting thing occurs:

$$K = 0 \text{ poles} = -1.5 \pm 0.5j \qquad \text{(damped oscillation)}$$
$$K = 10 \text{ poles} = -0.875 \pm 1.32j \qquad \text{(damped oscillation)}$$

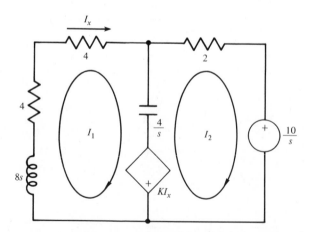

FIGURE 5-30. Transformed network of the network given in Fig. 5-29.

But when $K = 24$, $(3 - 0.125K) = 0!$ and therefore

$$\text{poles} = 0 \pm 1.58j \qquad \text{(no damping!)}$$

So, at a gain of 24, the circuit exhibits steady-state oscillation. For a gain greater than 24, you can show that the damping coefficient becomes positive, which is not physical for our model. This is an example of an *instability*. For gain greater than 24, the circuit is unstable and will not behave in a linear fashion.

Now let us consider the response for $K < 0$. In this case there is decaying oscillation until a critical value is reached, given by

$$(3 - 0.125K)^2 = 10$$

which gives $K = -1.298$. For this value there is no oscillation but a repeated pole at -3.16. The solution has the form

$$i(t) = -0.25 + 0.25e^{-3.16t} - 0.79te^{-3.16t}$$

For $K < -1.298$, the solution will be a dc level and decaying exponentials only. ∎

5-6

COMPUTER APPLICATIONS

You have already seen the many uses of the computer in solving network problems, including plotting, root finding, and solving sets of simultaneous equations. These form an important part of the practical approaches to solving real network problems. When an analysis requires that a Laplace transform be inverted to find the complete time function, including amplitudes, calculations by hand can become very cumbersome and therefore prone to error. There are a number of algorithms and programs available which will perform the inversion of a transform. In general, these routines require as input the coefficients of the polynomials making up the Laplace transform resulting from the analysis. In some cases the denominator must be factored first and the poles provided as input. The following example illustrates the use of computer routines to solve a problem in network analysis.

EXAMPLE 5-10
Find the complete time function of the Laplace transform given in Example 4-21.

SOLUTION
The poles of this transform were found in Example 4-21 to be 0, -2, -4, $4j$, and $-4j$. A computer program was used which requires as input the order of

the numerator polynomial and the coefficients in the numerator, the number of poles, and the values of these poles. The routine performs the partial fraction expansion and computes the values of the constants. This program reported the results shown in Fig. 5–31. Thus the time function is given by

$$v_{out}(t) = -0.52 - 14.7e^{-2t} + 23.2e^{-4t} - 8.1 \cos (4t) + 7.7 \sin (4t) \blacksquare$$

SUMMARY

This chapter presented a generalized method for solving any linear network using Laplace transforms in combination with the matrix methods of mesh and nodal analysis. To accomplish this goal, a number of specific topics were covered:

1. The initial condition voltage on a capacitor is expressed as either a dc voltage in series with the capacitor or an impulse current source shunting the capacitor.
2. The initial condition current through an inductor is expressed as either a dc current shunting the inductor or as an impulse voltage source in series with the inductor.
3. It was shown how resistors, capacitors, and inductors can be expressed as Laplace-transformed equivalent impedances or admittances by taking Laplace transform of their i–v relationships.
4. A network is transformed into s-space by the following steps:
   (a) The initial conditions are replaced by voltage or current sources.

```
PARTIAL FRACTION EXPANSION ROUTINE    INSTRUCTIONS?N
WHAT NUMERATOR ORDER?3

HOW MANY POLES (ORDER OF DENOMINATOR)?5
WHAT CONSTANT MULTIPLIER?1
INPUT 4     NUMERATOR COEFFICIENTS
A( 3     )?-33
A( 2     )?244
A( 1     )?0
A( 0     )?-66
INPUT   5      POLES AS (REAL,COMPLEX) NUMBERS.   THUS A POLE OF -6
WOULD BE INPUT AS (-6,0) AND ONE OF -5+4J  WOULD BE (-5,4)
P( 1     )?(0,0)
P( 2     )?(-2,0)
P( 3     )?(-4,0)
P( 4     )?(0,4)
P( 5     )?(0,-4)

THE AMPLITUDES ARE|
K( 1     ) =-.515625
K( 2     ) =-14.675
K( 3     ) = 23.2422
FOR THE COMPLEX POLE ( 0    , 4    ) THE COS AND SIN AMPLITUDES ARE
COS AMPLITUDE =-8.05156
SIN AMPLITUDE = 7.65469
```

FIGURE 5–31. Printout of a partial fraction expansion computer program run for Example 4-21.

(b) Resistance, capacitance, and inductance values are replaced by their transformed impedance or transformed admittance.

(c) All sources are replaced by their Laplace-transformed equivalent quantities.

(d) All controlled sources are replaced by their Laplace equivalent quantities using s-space dependent variables.

5. A set of rules were given which allow the matrix equation, in s-space, to be written directly using generalized mesh analysis.

6. A set of rules were given which allow the matrix equation, in s-space, to be written directly using generalized nodal analysis.

7. The form of the time dependence expected from some network solution was shown to be given only by the poles of a transform. These poles are the roots of the coefficient matrix determinant.

8. The final value theorem shows how the final value of a time function that does not oscillate steady state can be found by the value of $sF(s)$ as s goes to zero.

9. The initial value theorem shows how the initial value of a time function can be found by the value of $sF(s)$ as s tends infinity.

PROBLEMS

5–1   Prove that the same electrical characteristics are found for the inductor with an initial condition current, the dc current source model, and the impulse voltage source model.

5–2   Express the initial conditions of the components in Fig. 5–32 by voltage sources. Show a diagram and the values.

5–3   Express the initial conditions of the components in Fig. 5–32 by current sources. Show a diagram and the values.

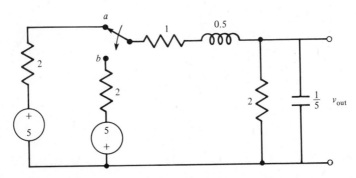

FIGURE 5–32. Network for Problems 5–2, 5–3, 5–4, 5–6, 5–7, and 5–9.

5–4   In the network of Fig. 5–32 the switch moves from position a to b at $t = 0$. Find the transformed networks, including initial conditions, for the generalized mesh approach (voltage sources) and the generalized nodal approach (current sources).

5–5   The switch in the network of Fig. 5–33 closes at $t = 0$. Find the transformed network suitable for generalized mesh analysis.

5–6   Develop the generalized mesh matrix equation for the network of Fig. 5–32 after the switch moves to position b.

5–7   Develop the generalized nodal matrix equation for the network of Fig. 5–32 after the switch moves to position b.

5–8   Develop the generalized mesh matrix equation for the network of Fig. 5–33 after the switch closes.

5–9   The switch in the network of Fig. 5–32 moves from position a to b at $t = 0$. Using generalized nodal analysis, obtain a complete solution for the output voltage. Plot the resulting solution and describe the results.

5–10   Find the output voltage for the network in Fig. 5–33 after the switch closes. Plot the result. Does the network exhibit gain? How long before all transients have died away?

5–11   Find the voltage across the capacitor in the network of Example 5-2.

5-12   Find the transformed matrix equation for the network in Fig. 5–34.

5–13   For the network of Fig. 5–35 assume that there are no initial conditions. Set up the transformed network and matrix equation for this network using nodal anal-

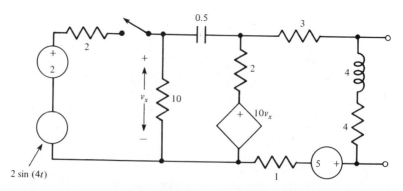

FIGURE 5–33. Network for Problems 5–5, 5–8, and 5–10.

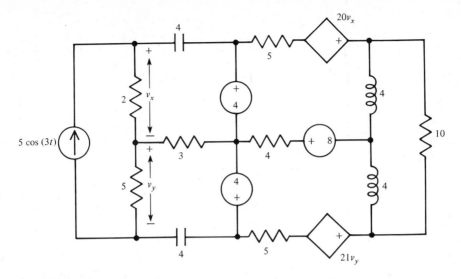

FIGURE 5–34. Network for Problem 5–12.

ysis and with a general frequency ω for the source. Find the amplitude of the output voltage for a general frequency. Plot this amplitude versus frequency from 0 to 10 rad/s. How would you describe the effect of this network in terms of frequency?

5–14 For the network of Fig. 5–36, assume that there are no initial conditions. Set up the transformed network and matrix equation for this network using mesh analysis for a general frequency ω of the source. Find the amplitude of the output voltage for a general frequency. Plot this amplitude versus frequency from 0 to 10 rad/s. How would you describe the effect of this network in terms of frequency? How does the performance of this network compare with that of Problem 5–13?

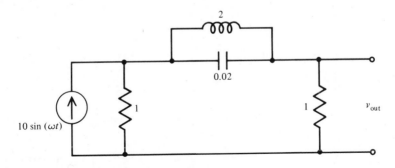

FIGURE 5–35. Network for Problem 5–13.

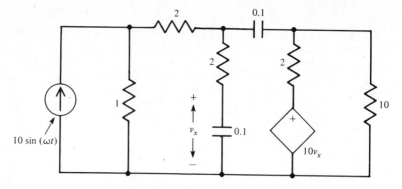

FIGURE 5–36.   Network for Problem 5–14.

5–15   The switch in the network of Fig. 5–37 closes at $t = 0$. Determine the form of the voltage $v_x(t)$ over positive values of K. Describe the type of time dependence for each appropriate range of K and give examples of the solution in each range.

COMPUTER PROBLEMS

5–16   Find and plot the time function of the Laplace transform

$$\frac{-42s^4 + 60s^3 - 80s + 44}{s^6 + 12s^5 + 58s^4 + 260s^3 + 504s^2 + 848s + 1152}$$

5–17   Find and plot the voltage of node 4 in the network of Fig. 5–21. Describe the characteristics of the resulting signal.

5–18   In the network of Fig. 5–37, suppose that there is a 5-Ω resistor shunting the 2-A source. Find the form of $v_x(t)$ as a function of K under this modification and compare with the results of Problem 5–15.

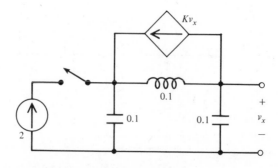

FIGURE 5–37.   Network for Problems 5–15 and 5–18.

CHAPTER 6

Fourier Analysis

Objectives

The global objective of this chapter is to learn the basic principles of Fourier series and transforms and to show how these can be used to evaluate the characteristics of electric signals. This is accomplished by satisfying the following specific objectives. At the conclusion of reading this chapter and doing the problems at the end of the chapter, you will be able to:

1. Define the Fourier series and its relationship to the time dependence of a signal.
2. Explain the relationship of time and level symmetry to the Fourier series of a time signal.
3. Calculate the Fourier series expansion of a periodic time function.
4. Apply the Fourier series expansion to problems in network analysis.
5. Define the complex form of the Fourier series.
6. Show how the Fourier transform is developed from the Fourier series and explain its significance.
7. Define the discrete Fourier transform and give examples of its application.

INTRODUCTION TO FOURIER SERIES

In this section the background ideas behind Fourier series analysis are given together with a formal definition of the series expansion. Examples of finding the Fourier series expansion of simple functions are given.

6-1.1 Concept of a Series Expansion

Before presenting the definition of a Fourier series expansion, it will be helpful to review the concept of series expansions in general.

Series Expansion.   Suppose that there is some function of x, given by $f(x)$, which is defined in an interval of x from a to b. This means that for every value of x within the interval there is a value of the function defined by $f(x)$. Often a graph of $f(x)$ versus x is used to describe the continuous variation of $f(x)$ with x. Figure 6–1 shows a hypothetical graph of some such function.

In some cases it is possible to write an equation for $f(x)$, and in others $f(x)$ may exist only as a smooth curve like Fig. 6–1 or even just a discrete set of function values at discrete values of x.

A *series expansion* is a means of writing an *equation* for $f(x)$ in terms of a set of other functions of x. To review this concept, let us consider one of the most common types: the *power series* expansion. In this case $f(x)$ is written in terms of the set of functions of x consisting of x raised to increasing positive powers. This would be written as

$$f(x) = \sum_{n=0}^{\infty} a_n x^n \qquad (6\text{-}1)$$

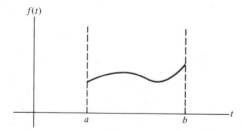

FIGURE 6–1.   Curve defined between two limits for which a series expansion may be desired.

where a_n = coefficient depending on n

$\sum$ = symbol meaning to sum terms for $n = 0, 1, 2, \ldots, n$, all the way to ∞.

If written out for the first few terms, Eq. (6-1) would have the form

$$f(x) = a_0 + a_1x + a_2x^2 + a_3x^3 + \cdots$$

The basic idea, then, is that for a given $f(x)$, values of the coefficients can be found such that Eq. (6-1) will reproduce the correct values of $f(x)$ for a given x. Thus, if such a series expansion were deduced for Fig. 6–1, a plot of $f(x)$ calculated from the series expansion would, in principle, *exactly* reproduce the same curve.

Accuracy. The degree to which the series expansion actually matches the function is dependent on how many terms in the series expansion are used in the calculation. In general, a series expansion involves an infinite number of terms and it is necessary to terminate the calculation at some point. The number of terms to include is determined by the accuracy desired in the matching of function and expansion. For example, if an accuracy of 1% is desired, terms would be included until the correction provided by including the next term was less than 1%. This is not a guarantee that the error will be less, since the accumulation of *all* the remaining terms may provide a correction exceeding 1%, but it is a good initial estimate of where to stop.

EXAMPLE 6-1

The common exponential function e^x has a power series expansion given by

$$e^x = 1 + x + \frac{x^2}{2!} + \frac{x^3}{3!} + \cdots + \frac{x^n}{n!} + \cdots$$

Find the number of terms that must be included to provide an accuracy of 0.5% for $x \leq 2$. What is the net error? Plot the function and the series expansion to third order for $x = 0$ to 3 and compare.

SOLUTION

The error will be greatest for the largest value of x for a series like this. Therefore, the number of terms will be found to make the error $< 0.5\%$ for $x = 2$. This is done by calculating the series and the error for each successive term.

$$n = 0 \qquad f = 0 \qquad \frac{2^0}{0!} = 1$$

$$n = 1 \qquad f = 1 \qquad \frac{2^1}{1!} = 2$$

$$n = 2 \qquad f = 3 \qquad \frac{2^2}{2!} = 2 \qquad E = 67\%$$

$$n = 3 \qquad f = 5 \qquad \frac{2^3}{3!} = 1.33333 \qquad E = 27\%$$

$$n = 4 \qquad f = 6.33333 \qquad \frac{2^4}{4!} = 0.66666 \qquad E = 10.5\%$$

$$n = 5 \qquad f = 7 \qquad \frac{2^5}{5!} = 0.26666 \qquad E = 3.8\%$$

$$n = 6 \qquad f = 7.26666 \qquad \frac{2^6}{6!} = 0.08888 \qquad E = 1.2\%$$

$$n = 7 \qquad f = 7.35555 \qquad \frac{2^7}{7!} = 0.02539 \qquad E = 0.3\%$$

Since the difference between the sixth and seventh terms is less than 0.5%, the series can be stopped at $n = 6$. The net error can be calculated since the value of e^2 can be determined from a calculator. The error is

$$E_{net} = \frac{e^2 - 7.35555}{e^2} = \frac{7.38905 - 7.35555}{7.38905}$$

$$= 0.45\%$$

So in this case even the net accuracy is within that required. This means that all of the succeeding terms ($n > 6$) do not contribute sufficient change to drive the error beyond 0.5%. Figure 6–2 shows the power series of $n = 3$ plotted from 0 to 5 and the function itself. You can see that considerable error occurs with such a limited number of terms in the series. ∎

The power series uses coefficients and powers of x as the basis on which to build the series. Many other bases are used to construct a series expansion of functions. For example, the exponential function is often used as the basis for constructing the series expansion of other functions. The Fourier series expansion uses the sine and cosine functions as the bases of a series expansion.

6-1.2 Fourier Series Expansion

The Fourier series expansion is most useful for constructing equations that describe functions which are *periodic* in time. Such functions have the property

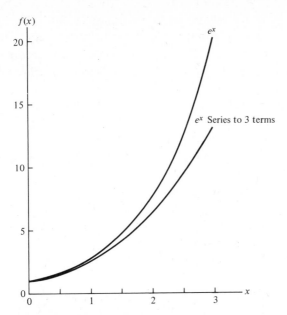

FIGURE 6–2. Comparison of the function e^x and a power series expansion to three terms from Example 6-1.

that they replicate themselves periodically in a fixed time span called the *period* of the function. If $f(t)$ describes the function and the period is given to be T, the periodicity is described by

$$f(t) = f(t + T) = f(t + 2T) \cdots$$

or, in general,

$$f(t) = f(t + nT) \tag{6-2}$$

where n is any integer.

Equation (6-2) would, in general, be valid for negative integers also, as long as the function is defined for $t < 0$. Figure 6–3 shows some such periodic

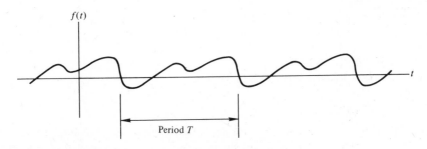

FIGURE 6–3. Periodic function with a period of *T* seconds.

function. It is typically not possible to find a simple equation that describes oddly varying functions such as this. The *objective* of Fourier series expansion is to construct an equation that will replicate, to within some stated accuracy, the variation $f(t)$ with t over all time for which the function is defined.

Definition. The formal definition of the Fourier series can be stated as follows:

Fourier Series Expansion. *Given a periodic function $f(t)$ with a period T and with the following conditions:*

1. *The function has a finite number of discontinuities within one period.*
2. *The function has a finite number of minima and maxima within one period, which are all of finite magnitude.*
3. *The function satisfies the integral relation*

$$\int_t^{t+T} |f(t)| \, dt < \infty$$

This means that the absolute area under the function curve over one period is finite.

Then the function can be expanded in a Fourier series defined by the equations

$$f(t) = a_0 + \sum_{n=1}^{\infty} [a_n \cos (n\omega_0 t) + b_n \sin (n\omega_0 t)] \qquad (6\text{-}3)$$

where the coefficients are defined by the integral relations

$$a_0 = \frac{1}{T} \int_0^T f(t) \, dt \qquad (6\text{-}4)$$

$$a_n = \frac{2}{T} \int_0^T f(t) \cos (n\omega_0 t) \, dt \qquad (6\text{-}5)$$

$$b_n = \frac{2}{T} \int_0^T f(t) \sin (n\omega_0 t) \, dt \qquad (6\text{-}6)$$

where $\omega_0 = 2\pi/T$ is called the fundamental *angular frequency. The fundamental frequency is given by the relation $f_0 = 1/T = \omega_0/2\pi$.*

Conditions. The conditions set down at the beginning of this definition do not present problems with the majority of periodic functions encountered in electrical and electronic networks. The conditions are necessary to assure that the various integrals used to find the coefficients can be performed. A discontinuity in a function occurs when the function displays a change in value at an instant of time. The square wave is a common example of such a function since it periodically changes from one level to another. In such a case the *value* of the function is *not defined* at the instant of the discontinuity but, rather, infinitesimally before and after its occurrence.

Integration Limits. To find the values of the coefficients the integrals of Eqs. (6-4), (6-5), and (6-6) must be evaluated for the specific $f(t)$ of the problem. The limits shown in these equations are from $t = 0$ to $t = T$, which is a span of one full period. It is important to realize that *any* limits of integration can be used as long as the limits span one full period. Thus the limits could just as well be from $-T/2$ to $+T/2$ or T to $2T$ or any span of one full period.

Integral Construction. There may be some confusion over the fact that the purpose of Fourier series expansion is to construct an equation for $f(t)$, and yet to do this the function must be integrated. How can this be done if an equation for the function is not already known? The answer is that it is often possible to write an equation for $f(t)$ *over one period*. In many cases, because of discontinuities, the equation will be written in several parts, as shown in the examples to follow. In any event, as long as an equation can be written over one full period, the integrals can be performed. The resulting Fourier series defined by Eq. (6-3) will be valid *over all time* and not just one period. In other cases the integrals may have to be performed numerically by a computer in order to determine the coefficients.

Harmonics. It is usually possible to evaluate the integrals of Eqs. (6-5) and (6-6) with the integer n left unspecified. The integration gives the coefficients as functions of n. In this way a coefficient for a particular term is found by simply substituting the value of n for that term into the equation for the coefficient. The number of terms in the series to be used for representing the function is again determined by the accuracy desired in the representation. Notice that the frequencies which are present in the Fourier expansion of a function are integer quantities of the fundamental frequency (i.e., ω_0, $2\omega_0$, $3\omega_0$, . . .). The individual terms are called the *harmonics* of the fundamental frequency. A Fourier series out to three harmonics would mean that for the required accuracy it was necessary to include terms from $n = 1$ to $n = 3$.

The following example shows how the Fourier series expansion of a given time signal is calculated using the relations of Eqs. (6-4), (6-5), and (6-6). Section 6-2 will show how much of the calculation required can often be eliminated by using observable symmetries of the function. When there are no symmetries, however, the rigorous performance of these integrations will always give the Fourier coefficients.

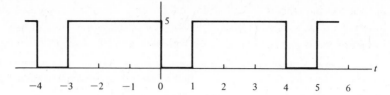

FIGURE 6–4. Signal for Example 6-2.

EXAMPLE 6-2

The time signal in Fig. 6–4 represents a negative pulse train. Find the Fourier series expansion of this signal. Plot the result for a series out to the seventh harmonic.

SOLUTION

The first step will be to determine the period and the fundamental angular frequency. Note that the signal begins to repeat itself after 4 s. Therefore, the period is $T = 4$ and the angular frequency is $\omega_0 = 2\pi/T = 2\pi/4 = \pi/2$ rad/s. The second step will be to determine the best limits of integration spanning one full period and then to express the function $f(t)$ over that span. Let us just pick limits from 0 to 4. Then from Fig. 6–4 it is clear that the function can be written

$$f(t) = \begin{cases} 0 & 0 < t < 1 \\ 5 & 1 < t < 4 \end{cases}$$

This means that the integrals need only extend from 1 to 4 s since the function is zero from 0 to 1. The dc level is found from Eq. (6-4):

$$a_0 = \frac{1}{T} \int_0^T f(t) \, dt = \frac{1}{4} \int_1^4 5 \, dt = \frac{5}{4}(4 - 1) = \frac{15}{4}$$

The cosine coefficient is found from Eq. (6-5):

$$a_n = \frac{2}{T} \int_0^T f(t) \cos(n\omega_0 t) \, dt = \frac{2}{4} \int_1^4 5 \cos\left(\frac{n\pi t}{2}\right) dt$$

$$= \frac{5}{n\pi} \int_1^4 \cos\left(\frac{n\pi t}{2}\right) d\left(\frac{n\pi t}{2}\right) = \frac{5}{n\pi} \sin\left(\frac{n\pi t}{2}\right)\Bigg]_1^4$$

$$= \frac{5}{n\pi} \left[\sin(2n\pi) - \sin\left(\frac{n\pi}{2}\right) \right]$$

$$= -\frac{5}{n\pi} \sin\left(\frac{n\pi}{2}\right)$$

272

The sine coefficient is found from Eq. (6-6):

$$b_n = \frac{2}{T} \int_0^T f(t) \sin(n\omega_0 t)\, dt = \frac{2}{4} \int_1^4 5 \sin\left(\frac{n\pi t}{2}\right) dt$$

$$= \frac{5}{n\pi} \int_1^4 \sin\left(\frac{n\pi t}{2}\right) d\left(\frac{n\pi t}{2}\right) = \left. -\frac{5}{n\pi} \sin\left(\frac{n\pi t}{2}\right) \right]_1^4$$

$$= -\frac{5}{n\pi}\left[\cos(2n\pi) - \cos\left(\frac{n\pi}{2}\right)\right]$$

$$= -\frac{5}{n\pi}\left[1 - \cos\left(\frac{n\pi}{2}\right)\right]$$

The Fourier series expansion can be constructed by simply substituting values of n into the equations for a_n and b_n. The problem specifies out to the seventh harmonic, $n = 7$:

$n = 1$	$a_1 = -\dfrac{5}{\pi}$	$b_1 = -\dfrac{5}{\pi}$
$n = 2$	$a_2 = 0$	$b_2 = -\dfrac{5}{\pi}$
$n = 3$	$a_3 = \dfrac{5}{3\pi}$	$b_3 = -\dfrac{5}{3\pi}$
$n = 4$	$a_4 = 0$	$b_4 = 0$
$n = 5$	$a_5 = -\dfrac{5}{5\pi}$	$b_5 = -\dfrac{5}{5\pi}$
$n = 6$	$a_6 = 0$	$b_6 = -\dfrac{5}{3\pi}$
$n = 7$	$a_7 = \dfrac{5}{7\pi}$	$b_7 = -\dfrac{5}{7\pi}$

Now the time function can be written out (to seven harmonics) using Eq. (6-3) as a guide:

$$f(t) = \frac{15}{4} - \left(\frac{5}{\pi}\right)\left[\cos\left(\frac{\pi t}{2}\right) + \sin\left(\frac{\pi t}{2}\right) + \sin(\pi t) - \frac{\cos(3\pi t/2)}{3}\right.$$

$$+ \frac{\sin(3\pi t/2)}{3} + \frac{\cos(5\pi t/2)}{5} + \frac{\sin(5\pi t/2)}{5} + \frac{\sin(3\pi t)}{3}$$

$$\left. - \frac{\cos(7\pi t/2)}{7} + \frac{\sin(7\pi t/2)}{7}\right]$$

This function is plotted in Fig. 6–5. You will note that whereas the shape is definitely beginning to look like Fig. 6–4, there is still considerable error. This is usually the case for functions with sharp discontinuities. In Section 6-2 examples of other signals are given where even three harmonics comes very close to replicating the original function. ∎

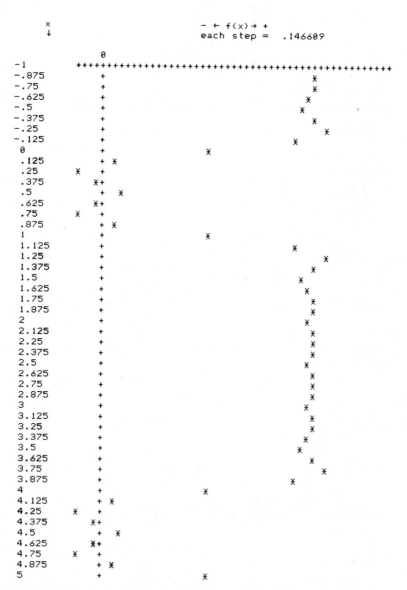

FIGURE 6–5. Plot of the Fourier series of the signal of Fig. 6–4 out to seven harmonics.

6-1.3 Interpretation of Fourier Series

The preceding section presented the basic definition of Fourier series expansion of periodic functions. The ultimate result of such an expansion is the representation of a periodic function by an equation in sine and cosine functions as given by Eq. (6-3). The ability to write a single equation that describes arbitrary periodic functions is often very useful, but this is not the only, or even primary usefulness of Fourier series.

The Fourier series expansion of a periodic function tells us about the *frequency* content of the signal. This means the extent to which the signal depends on high-frequency (fast) variation in time and lower-frequency (slower) variation in time. Such a description is called a *spectrum analysis* of the signal. The lowest frequency of dependence is the fundamental, ω_0. The highest is, in principle, unlimited since the sum extends to an infinite harmonic but is limited practically when the value of the coefficients become very small. Spectrum analysis means a study of the coefficient magnitudes as a function of harmonic frequency. In order to study the frequency content (i.e., the spectrum), it is convenient to use a slightly different representation of the series expansion.

Cosine Representation. In some cases it is inconvenient to have the Fourier series expressed as the sum of separate sine and cosine functions. This can be rectified by using the trigonometric property that

$$A \sin (\omega t) + B \cos (\omega t) = C \cos (\omega t + \phi)$$

where

$$C = (A^2 + B^2)^{1/2}$$

$$\phi = -\tan^{-1}\left(\frac{A}{B}\right)$$

When this property is used with the definition of Eq. (6-3), it becomes possible to write the Fourier series in the form

$$f(t) = c_0 + \sum_{n=1}^{\infty} c_n \cos (n\omega_0 t + \phi_n) \qquad (6\text{-}7)$$

where

$$c_0 = a_0 \qquad (6\text{-}8)$$

$$c_n = [(a_n)^2 + (b_n)^2]^{1/2} \qquad (6\text{-}9)$$

$$\phi_n = -\tan^{-1}\left(\frac{b_n}{a_n}\right) \qquad (6\text{-}10)$$

The same integrals for coefficients will be evaluated as given in Eqs. (6-4), (6-5), and (6-6) when using this representation of the Fourier series. Once these integrals have been evaluated, the equations above can be used to set up the series expansion in this form. This is often called the *magnitude and phase representation* of the Fourier series.

The advantage of this representation is that the c_n are net amplitudes of each frequency harmonic, $n\omega_0$. Thus these amplitudes represent the spectrum analysis since a comparison of the values for different frequencies tells the extent to which the signal *depends* on these frequencies.

Phase. Great care must be taken when computing the phase by Eq. (6-10) to be sure the resulting phase angle is in the correct quadrant. Consider the diagram of Fig. 6–6 to see why. In the first case suppose that $a = 4$ and $b = 4$ (first quadrant); then the phase would be $-\phi = \tan^{-1}(4/4) = \tan^{-1}(1) = 45°$ (using a calculator). Now suppose that $a = 4$ and $b = -4$ (fourth quadrant), then $-\phi = \tan^{-1}(-4/4) = \tan^{-1}(-1) = -45°$, which is correct. Now take the case of $a = -4$ and $b = 4$ (third quadrant); then the phase is $-\phi = \tan^{-1}(4/4) = \tan^{-1}(-1) = -45°$ (by calculator). But this is not correct! The correct phase is $135°$. The calculator, and even computer algorithms for computing $\tan^{-1}(x)$ cannot distinguish the first and third or the second and fourth quadrant. When calculating the phase by Eq. (6-10) it is necessary to make the necessary corrections by $90°$ increments to correct for such quadrant error.

Line Spectrum. A plot of the amplitudes in the cosine representation, Eq. (6-7), versus the harmonic frequencies will consist of lines at each harmonic

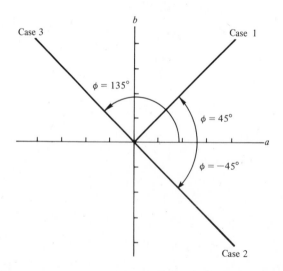

FIGURE 6–6. Phase relations in the cosine representations can confuse quadrants.

whose height is equal to the amplitude of that harmonic. Such a plot is called the *line spectrum* of the signal and shows the relative extent to which each frequency contributes to the full signal. The phase is also represented by such a line plot.

EXAMPLE 6-3

Calculate the cosine Fourier series representation of the signal in Fig. 6–4 and show a line spectrum out to the 16th harmonic.

SOLUTION

The cosine representation can be easily found since the a_n and b_n have been found in Example 6-2. A computer program was written to calculate the c_n out to $n = 16$, but let us do one by hand to see how it goes.

$$n = 1 \quad c_1 = \left[\left(\frac{5}{\pi} \right)^2 + \left(\frac{5}{\pi} \right)^2 \right]^{1/2} = 2.25079$$

$$\phi_1 = -\tan^{-1} \left(\frac{-5/\pi}{-5/\pi} \right) = -\tan^{-1}(1)$$

$$= -(45°)$$

but this is third quadrant! So

$$\phi_1 = -(45° + 180°) = -225° = 135°$$

The results of the simple computer program to do this gives (angles in radians)

n	c_n	ϕ_n
0	3.75	0
1	2.251	2.356
2	1.592	1.571
3	0.750	0.785
4	0	0
5	0.450	2.356
6	0.531	1.571
7	0.322	0.785
8	0	0
9	0.250	2.356
10	0.318	1.571
11	0.205	0.785
12	0	0
13	0.173	2.356
14	0.227	1.571
15	0.150	0.785
16	0	0

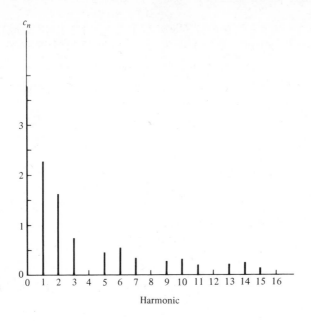

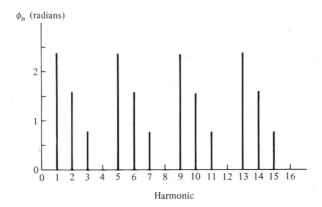

FIGURE 6–7. Amplitude and phase line spectrum of the Fourier series cosine representation of the signal in Fig. 6–4.

The graphs of Fig. 6–7 shows the line spectrum of amplitude and phase for this signal. ∎

Power Spectrum. Another type of spectrum analysis which is often used is the contribution of the different harmonics to the net *power* of the signal. To construct such a spectrum, the assumption is made that the signal represents a voltage or current which is inserted into a *resistive* load. Then the power is simply *proportional* to the voltage or current squared. Thus the power spectrum is simply the square of the cosine-representation amplitude terms.

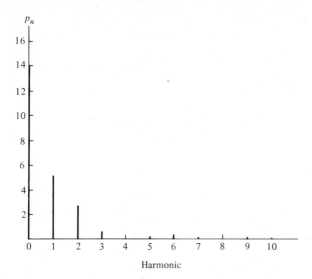

FIGURE 6–8. Power spectrum of the signal in Fig. 6–4.

$$p_n = (c_n)^2 \qquad (6\text{-}11)$$

where p_n is the power content of the nth harmonic.

Figure 6–8 shows the power spectrum for the signal of Fig. 6–4. You will see that a major part of the power delivered is contained in the dc term and the first three harmonics. This means that if a filter drastically cut down the fifth and higher harmonics, the power delivery would not be greatly affected, although the shape of the signal would be considerably modified.

6-2

FOURIER SERIES COMPUTATION

The preceding section defined the Fourier series in terms of the relation of Eq. (6-3) and the associated integrals for evaluation of the coefficients. An example showed how a series could be constructed from a given waveform. In this section a number of techniques are presented which can make determination of the coefficients easier. These techniques are based on an observation of certain *symmetries* about the function with respect to level and time.

6-2.1 Functional Symmetries

There are two types of symmetry which a time function can have, both of which will be helpful in finding the Fourier series expansion of the function. One of the symmetries has to do with reflections of the function above and below the zero signal level axis and the other with reflections of the function across the zero time axis.

Signal Level Area Symmetry.   The most important type of level symmetry from a Fourier series point of view involves the question of signal area above and below the zero level axis, over one full period. You can see from Eq. (6-4) that the dc term in the series is simply the *net area* under the signal curve over one period, divided by the period:

$$a_0 = \frac{\text{net area in one period}}{T} \tag{6-12}$$

Thus, if the area above the zero level axis is the same as the area below the zero axis, $a_0 = 0$ since the *net* area will be zero. Furthermore, it is often much easier to find the net area from geometric principles than to do the integral of Eq. (6-4) in order to find the a_0 term when it is not zero.

Signal Level Shape Symmetry.   In some cases a signal will have the same shape for positive signal level and negative signal level, over one period. Often, the negative signal shape is a mirror image or reverse mirror image of the positive signal shape. In either case it is obvious that the net area will be zero and thus that $a_0 = 0$.

EXAMPLE 6-4
Use symmetry and geometric considerations to find the dc term of a series expansion for the three signals in Fig. 6–9.

SOLUTION
Signal (a): The period is quickly seen to be 5 s. By geometrical considerations the area above is the sum of the areas of two rectangles:

$$\text{area above} = (1)(1) + (2)(1) = 3$$

The area above is the area of a single rectangle:

$$\text{area below} = (-1)(3) = -3$$

Since the area above and below are the same, the dc term will be zero since the net area is zero; thus $a_0 = 0$.
   Signal (b): This signal clearly has signal level symmetry via a reversed mirror image over one period. Thus $a_0 = 0$. This can, of course, be verified by geometrical computation. There are triangles above and below.

$$\text{area above} = \tfrac{1}{2}(1)(10) = 5$$
$$\text{area below} = \tfrac{1}{2}(1)(-10) = -5$$

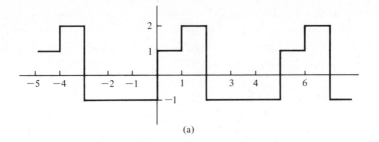

(a)

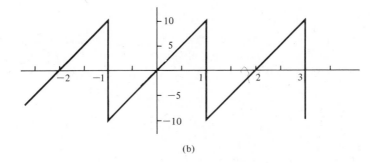

(b)

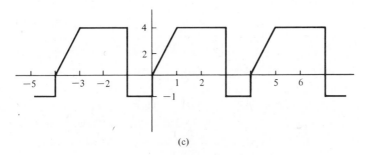

(c)

FIGURE 6–9.   Signals for Examples 6-4 and 6-5.

Hence the net area is zero and $a_0 = 0$.

Signal (c): This kind of signal is a good example of one for which geometrical computation of a_0 is much easier than integration. The period is 4 s.

$$\text{area above} = \text{sum of triangle} + \text{rectangle areas}$$

$$= \tfrac{1}{2}(1)(2) + (2)(2) = 5$$

$$\text{area below} = \text{rectangle area} = (1)(-1) = -1$$

By Eq. (6-12), a_0 is

$$a_0 = \frac{5 - 1}{4} = \frac{4}{4} = 1$$

∎

Even/Odd Time Symmetry. Symmetries in time have to do with the relationship between the value of a function at some time, t, compared to the value of that function at the same *negative* time, $-t$. There are three possibilities:

1. Even Symmetry. In some cases the value of a function at positive times t is exactly the same as the value of the function at negative times $-t$. Such functions are said to have *even symmetry*. In equation form this is written

$$f(t) = f(-t) \tag{6-13}$$

2. Odd Symmetry. In some cases the value of a function at negative times $-t$ is the same magnitude but the negative of the values of the function at positive times t. Such functions are said to have *odd symmetry*. In equation form this is written

$$f(t) = -f(-t) \tag{6-14}$$

3. No Symmetry. It is important to remember that not all functions have even or odd symmetry. For these functions there is no relationship between values of the function at positive times and values of the function at negative times.

Equation (6-13) must be found to hold for *all* times before the function can be said to have even symmetry. Another way of viewing even symmetry is to say that the function has its exact mirror image reflected across the $t = 0$ axis. Consider the function given in Fig. 6–10. This function has even symmetry. Notice that if the graph were to be folded at the $t = 0$ axis line, then when brought together the curves for $+t$ and $-t$ would match exactly. Note that $\cos (x)$ is an even function also, since $\cos (x) = \cos (-x)$. It can be shown that functions with even symmetry will have no terms in $\sin (n\omega_0 t)$ (i.e., $b_n = 0$). Thus whenever a function is found to have even symmetry it

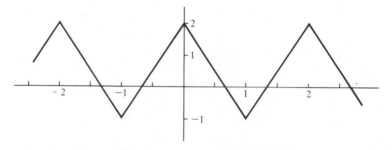

FIGURE 6–10. Signal with even time symmetry.

will not be necessary to compute the b_n integral since it will turn out to be zero for all cases.

If $f(t)$ is even, then $b_n = 0$ for all n

Equation 6-14 must be found to hold for *all* times before the function can be said to have odd symmetry. Odd symmetry is reverse mirror image across the $t = 0$ axis. The function of Fig. 6–9b has odd symmetry. Notice that for all values of t the value of the function has the opposite polarity of the function at $-t$. As usual the value of the function at a discontinuity is not considered, so the value at $t = 1$ and $t = -1$ are not considered. Notice that for t slightly less than $+1$ the function is almost 10, and for t slightly more positive than -1 the function is almost -10. If the graph is folded along the $t = 0$ axis and brought together, the curves will be equal but opposite. The sin (x) function is odd since sin $(x) = -\sin(-x)$. It can be shown that functions with odd symmetry will have no terms in cos $(n\omega_0 t)$ (i.e., $a_n = 0$ for all n). Thus it will be unnecessary to compute the integrals for a_n for odd functions since they will be zero.

If $f(t)$ is odd, then $a_n = 0$ for all n

EXAMPLE 6-5
Find the Fourier series of the signal given in Fig. 6–9b. Write the series out to three harmonics.

SOLUTION
Well, we have already seen that the dc level is zero by the level symmetry. In addition, the function is clearly of odd time symmetry. Thus

$$a_0 = 0$$

$$a_n = 0$$

This leaves the terms for b_n to be computed. The period is seen to be 2 s. Now to perform the integration of Eq. (6-6), it will be necessary to write an expression for the signal within one period. If we use the period of 0 to 2 s, the function, and integration, will be carried out in two parts, 0 to 1 and 1 to 2, since the function is different for the two spans. Notice, however, that from -1 to $+1$, which also covers one full period, the function has one form and therefore there will be only one integral. This is very important. The problem should be studied to find the integration limits, spanning one full period, which result in the easiest integration. The form of $f(t)$ is a straight-line equation with a zero intercept and having the value of 10 at 1 s,

$$f(t) = 10t \qquad -1 < t < +1$$

Then the integral for b_n is

$$b_n = \frac{2}{2} \int_{-1}^{+1} 10t \sin(n\pi t) \, dt = \frac{10}{(n\pi)^2} \int_{-1}^{+1} (n\pi t) \sin(n\pi t) \, d(n\pi t)$$

From an integral table it can be found that

$$\int x \sin(x) \, dx = \sin(x) - x \cos(x)$$

This allows the integral above to be written

$$b_n = \frac{10}{(n\pi)^2} [\sin(n\pi t) - (n\pi t) \cos(n\pi t)] \Big]_{-1}^{+1}$$

$$= -\frac{20}{n\pi} \cos(n\pi)$$

To write down the Fourier series to three terms means to use $n = 1, 2,$ and 3. The corresponding coefficients are found to be $b_1 = -(20/\pi) \cos(\pi) = 20/\pi$, $b_2 = -(20/2\pi) \cos(2\pi) = -10/\pi$, and $b_3 = -(20/3\pi) \cos(3\pi) = 20/3\pi$. The Fourier series becomes

$$f(t) = \left(\frac{20}{\pi}\right) \sin(\pi t) - \left(\frac{10}{\pi}\right) \sin(2\pi t) + \left(\frac{20}{3\pi}\right) \sin(3\pi t)$$

Figure 6–11 shows a plot of this function illustrating the degree to which three harmonics replicates the original signal. ∎

EXAMPLE 6-6

Find the Fourier series expansion to three harmonics of the signal shown in Fig. 6–10.

SOLUTION

The period of this signal is 2 s, so the angular frequency is $\omega_0 = \pi$ rad/s. The dc term can be found from geometrical considerations as follows. The line from $t = 0$ to $t = 1$ has the equation $f(t) = mt + b$ (i.e., a straight line). The

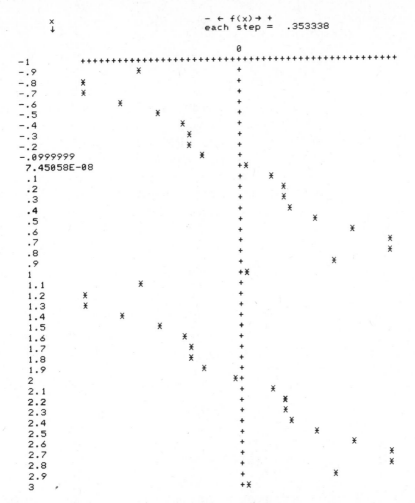

```
        x                    - ← f(x) → +
        ↓                    each step =  .353338


                                      0
  -1        +++++++++++++++++++++++++++++++++++++++++++++++++++++
  -.9                  ×                 +
  -.8        ×                           +
  -.7        ×                           +
  -.6            ×                        +
  -.5               ×                     +
  -.4                   ×                 +
  -.3                   ×                 +
  -.2                   ×                 +
  -.0999999                  ×           +
   7.45058E-08                           +×
   .1                                    +     ×
   .2                                    +       ×
   .3                                    +       ×
   .4                                    +      ×
   .5                                    +         ×
   .6                                    +           ×
   .7                                    +              ×
   .8                                    +              ×
   .9                                    +          ×
  1                                      +×
  1.1                  ×                 +
  1.2        ×                           +
  1.3        ×                           +
  1.4            ×                        +
  1.5               ×                     +
  1.6                 ×                   +
  1.7                   ×                 +
  1.8                   ×                 +
  1.9                     ×               +
  2                                      ×+
  2.1                                    +    ×
  2.2                                    +      ×
  2.3                                    +      ×
  2.4                                    +       ×
  2.5                                    +         ×
  2.6                                    +           ×
  2.7                                    +              ×
  2.8                                    +              ×
  2.9                                    +          ×
  3        ′                             +×
```

FIGURE 6—11. Plot of the Fourier series of the odd signal in Fig. 6—9b, to three harmonics.

intercept is just $b = 2$, since when $t = 0$ the function is at 2. The slope m is such that the function changes by -3 (2 to -1) while the time changes by only 1 s; thus $m = -3/1 = -3$. So the equation is $f(t) = -3t + 2$. You can verify that this equation matches the line. Now, this means that $f(t) = 0$ occurs when $0 = -3t + 2$ or at $t = \frac{2}{3}$ s. The area is

$$\text{area above} = \text{area of triangle of base } \tfrac{2}{3} \text{ and height } 2$$

$$= \tfrac{1}{2}(\tfrac{2}{3})(2) = \tfrac{4}{6}$$

$$\text{area below} = \text{area of triangle of base } \tfrac{1}{3} \text{ and height } -1$$

$$= \tfrac{1}{2}(\tfrac{1}{3})(-1) = -\tfrac{1}{6}$$

So the dc term can be found from Eq. (6-12) as

$$a_0 = \frac{2(\frac{4}{6} - \frac{1}{6})}{2} = \frac{1}{2} = 0.5$$

The reason for the 2 is that we found the area of only half the period, but it is clear that the same area relationship holds for the other half as well, so the result can just be doubled. It was deduced earlier that his function has even symmetry and therefore that all $b_n = 0$. So to find the a_n it will be necessary to set up the integral of Eq. (6-5) over one full period. Instead of 0 to 2, let us use -1 to $+1$. The equation for 0 to 1 was just found to be $f(t) = -3t + 2$. You can see that for -1 to 0 the equation simply has opposite slope; thus $f(t) = 3t + 2$. This gives

$$a_n = \frac{2}{2} \int_{-1}^{0} (3t + 2) \cos(n\pi t) \, dt + \frac{2}{2} \int_{0}^{+1} (-3t + 2) \cos(n\pi t) \, dt$$

Notice that if the variable of integration in the first integral is changed to $-t$, the entire term will become equal to the second integral if the limits of integration are also reversed (remember that this changes the sign of the integral). So the equation for a_n becomes

$$a_n = 2 \int_{0}^{1} (-3t + 2) \cos(n\pi t) \, dt$$

$$= -6 \int_{0}^{1} t \cos(n\pi t) \, dt + 4 \int_{0}^{1} \cos(n\pi t) \, dt$$

You can verify that the result of performing these integrals gives

$$a_n = \frac{-6}{(n\pi)^2} [\cos(n\pi) - 1]$$

Writing the first three harmonics would mean the first three terms. Notice that for n even, $\cos(n\pi) = +1$ and the coefficient is zero. Thus the first three nonzero harmonics will be for $n = 1, 3,$ and 5.

$$a_1 = -\left(\frac{6}{\pi^2}\right)(-1 - 1) = \frac{12}{\pi^2}$$

$$a_3 = -\left(\frac{6}{9\pi^2}\right)(-1 - 1) = \frac{12}{9\pi^2}$$

$$a_5 = -\left(\frac{6}{25\pi^2}\right)(-1 - 1) = \frac{12}{25\pi^2}$$

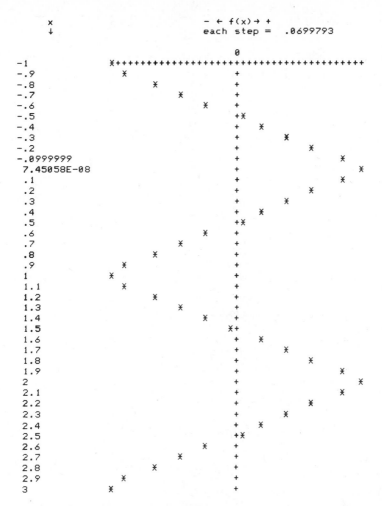

```
                  x                      -  ← f(x) → +
                  ↓                      each step =  .0699793

                                                 0
           -1           ✗+++++++++++++++++++++++++++++++++++++++++++++++++++++++
           -.9              ✗                    +
           -.8                   ✗               +
           -.7                        ✗          +
           -.6                           ✗       +
           -.5                                   +✗
           -.4                                   +    ✗
           -.3                                   +       ✗
           -.2                                   +          ✗
           -.0999999                             +             ✗
            7.45058E-08                          +                ✗
            .1                                   +             ✗
            .2                                   +          ✗
            .3                                   +       ✗
            .4                                   +    ✗
            .5                                   +✗
            .6                              ✗    +
            .7                           ✗       +
            .8                      ✗            +
            .9                  ✗               +
           1              ✗                     +
           1.1               ✗                  +
           1.2                   ✗               +
           1.3                        ✗          +
           1.4                           ✗       +
           1.5                                   ✗+
           1.6                                   +    ✗
           1.7                                   +       ✗
           1.8                                   +          ✗
           1.9                                   +             ✗
           2                                     +                ✗
           2.1                                   +             ✗
           2.2                                   +          ✗
           2.3                                   +       ✗
           2.4                                   +    ✗
           2.5                                   +✗
           2.6                              ✗    +
           2.7                           ✗       +
           2.8                      ✗            +
           2.9                  ✗               +
           3              ✗                     +
```

FIGURE 6–12.   Plot of the Fourier series of the even signal in Fig. 6–10, to three nonzero harmonics.

So the Fourier series is

$$f(t) = \frac{12}{\pi^2} \left[\cos (\pi t) + \frac{1}{9} \cos (3\pi t) + \frac{1}{25} \cos (5\pi t) \right]$$

This function is plotted in Fig. 6–12.   ∎

Half-Wave Symmetry.   Another type of symmetry which is of value in computing the Fourier series amplitudes has to do with the relationship between the function value at t and the value of the function at $t + T/2$ (i.e., one half period later). One possibility is that $f(t) = +f(t + T/2)$, but this is just the definition of a periodic function with a period of $T/2$! In other words, the period is not T it is $T/2$. The more interesting case is those functions with a definite period of T but where there is an odd relationship, $f(t) = -f(t + T/2)$. This is the definition of half-wave symmetry. Of course, the last possibility is that there is no relation between the value at t and the value at $t + T/2$. It turns out that functions with half-wave symmetry have only odd harmonics:

Half-wave symmetric functions *have* $a_n = 0$ *and* $b_n = 0$ *for even harmonics of* $n = 2, 4, 6, 8, \ldots$.

EXAMPLE 6-7
Show that the function of Fig. 6–13 has half-wave symmetry. Find the Fourier series amplitudes and show that the even harmonic amplitudes are zero.

SOLUTION
First note that the period is $T = 4$ s. Now for the symmetry. Well, $f(1) = 4$ and $f(-1) = -4$, so it almost looks odd. But then you can see that this is only a singular point where the odd-function condition is satisfied. For example, $f(0.5) = 2$, but $f(-0.5) = -6$ so it cannot be odd. It does have half-wave symmetry, however. For example, $f(1 + 4/2) = f(3) = -4 = -f(1)$ and $f(0.5 + 2) = -2$, which is $-f(0.5)$. In general, you can see that $f(t) = -f(t + 2)$.

Now to find the Fourier series amplitudes. The dc term is zero by inspection since the area above and the area below are clearly equal, $a_0 = 0$. Let us integrate from 0 to 4 directly; then

$$f(t) = \begin{cases} 4t & 0 < t < 2 \\ -4t + 8 & 2 < t < 4 \end{cases}$$

The a_n term is found as follows:

$$a_n = \frac{1}{2} \int_0^2 4t \cos\left(\frac{n\pi t}{2}\right) dt + \frac{1}{2} \int_2^4 (-4t + 8) \cos\left(\frac{n\pi t}{2}\right) dt$$

$$= \frac{8}{(n\pi)^2} \left[\cos\left(\frac{n\pi t}{2}\right) + \frac{n\pi t}{2} \sin\left(\frac{n\pi t}{2}\right) \right]_0^2$$

$$- \frac{8}{(n\pi)^2} \left[\cos\left(\frac{n\pi t}{2}\right) + \frac{n\pi t}{2} \sin\left(\frac{n\pi t}{2}\right) \right]_2^4$$

$$+ \frac{8}{n\pi} \sin\left(\frac{n\pi t}{2}\right) \Big]_2^4$$

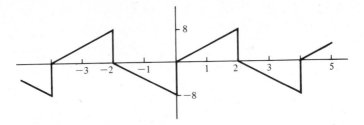

FIGURE 6-13. Signal for Example 6-7.

When the limits are substituted and the expression simplified, the amplitude turns out to be

$$a_n = \frac{16}{(n\pi)^2} [\cos(n\pi) - 1]$$

Now, for n even, $\cos(n\pi) = +1$ and thus $a_n = 0$, as the half-wave symmetry predicts. For n odd, $\cos(n\pi) = -1$ and $a_n = -32/(n\pi)^2$.

The b_n coefficients are evaluated by similar integrals, which give the result,

$$b_n = \frac{8}{n\pi} [1 - \cos(n\pi)]$$

In this case again, $b_n = 0$ for even values of n, while for odd values, $b_n = 16/n\pi$. The first few terms of the series expansion are

$$f(t) = -\frac{32}{\pi^2} \cos\left(\frac{\pi t}{2}\right) + \frac{16}{\pi} \sin\left(\frac{\pi t}{2}\right) - \frac{32}{9\pi^2} \cos\left(\frac{3\pi t}{2}\right) + \frac{16}{3\pi} \sin\left(\frac{3\pi t}{2}\right)$$

This function is plotted in Fig. 6–14 for the fifth harmonic. ■

6-2.2 Calculational Techniques

The purpose of this section is to present a summary of the calculational techniques outlined in previous sections for finding the Fourier series expansion of a periodic function. In addition, computer usage in Fourier series analysis is discussed.

Summary of Fourier Series Construction. The construction of the Fourier series expansion of a periodic function can be accomplished by the following sequence of operations:

1. Determine the period of the function, T. Calculate the fundamental angular frequency, $\omega_0 = 2\pi/T$.

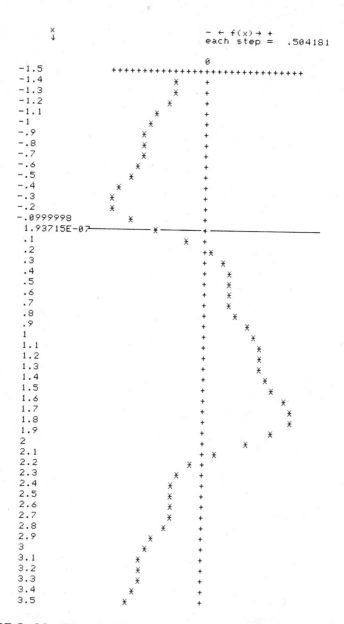

FIGURE 6–14. Plot of the Fourier series of the signal of Fig. 6–13, with half-wave symmetry.

2. Examine the function for any symmetries in level or time. From the symmetry, if any, set appropriate harmonic amplitudes to zero.
3. If not zero by symmetry, calculate the dc term, a_0, by means of geometrical construction or the integral of Eq. (6-4).
4. Select limits for harmonic amplitude integration which will give the easiest integrals. Determine the equation for the function within these limits.
5. Perform the integrations for any amplitudes not zero by symmetry using Eq. (6-5) and/or Eq. (6-6).

The following example illustrates application of these operations.

EXAMPLE 6-8
Find the Fourier series expansion of the half-wave rectified signal of Fig. 6–15. Plot the resulting series for the first four nonzero harmonics.

SOLUTION
Let us just follow the operations outlined above.
1. The period is evidently $T = 2$ s since the function repeats itself every 2 s. Then the fundamental frequency is $\omega_0 = 2\pi/2 = \pi$ rad/s.
2. There is no level symmetry since the function never goes negative. However, you can see that the function does have *even* time symmetry, since $f(t)$ is equal to $f(-t)$. Note that if $f(t) = 0$ and if $f(-t) = 0$ also, this can satisfy either even or odd symmetry. The rest of the function determines the symmetry. Therefore, all $b_n = 0$ and it will not be necessary to perform integration for this term.
3. It will be necessary to determine the dc term by integration. To do this the function will have to be defined, as operation 4 and the limits of integration. If the integration is from -0.5 to 1.5, the function takes on a simple form. From 0.5 to 1.5 it is clearly zero. From -0.5 to $+0.5$ it is one half cycle of a cosine wave. The period of this wave is also 2 s, and the amplitude is 5. Thus

$$f(t) = \begin{cases} 5 \cos (\pi t) & -0.5 < t < +0.5 \\ 0 & 0.5 < t < 1.5 \end{cases}$$

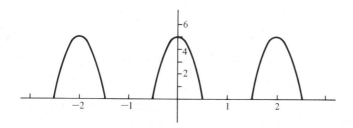

FIGURE 6–15. Half-wave rectified signal for Example 6-8.

The dc term becomes

$$a_0 = \tfrac{1}{2} \int_{-1/2}^{1/2} 5 \cos (\pi t) \, dt = \left. \frac{5}{2\pi} \sin (\pi t) \right]_{-1/2}^{1/2}$$

$$= \frac{5}{\pi}$$

For a_n the integral is

$$a_n = \frac{5}{\pi} \int_{-1/2}^{1/2} \cos (\pi t) \cos (n\pi t) \, d(\pi t)$$

A book of integral tables gives

$$\int \cos (mx) \cos (nx) \, dx = \frac{\sin [(m - n)x]}{2(m - n)} + \frac{\sin [(m + n)x]}{2(m + n)}$$

where in our integral $m = 1$, $n = n$, and $x = \pi t$. No problem, except that the table also says $n^2 \neq m^2$. But when $n = 1$ this condition will be true. Then we must use the integral

$$\int \cos^2 (x) \, dx = \frac{x}{2} + \frac{\sin (2x)}{4}$$

So a_1 is found from this last integral:

$$a_1 = \frac{5}{\pi} \int_{-1/2}^{1/2} \cos^2 (\pi t) \, d(\pi t)$$

$$= \frac{5}{\pi} \left[\frac{\pi t}{2} + \frac{\sin (2\pi t)}{4} \right]_{-1/2}^{1/2} = 2.5$$

The rest of the coefficients are found as

$$a_n = \frac{5}{\pi} \left[\frac{\sin [\pi t(1 - n)]}{2(1 - n)} + \frac{\sin [\pi t(1 + n)]}{2(1 + n)} \right]_{-1/2}^{1/2}$$

$$= \frac{5}{\pi} \left[\frac{\sin [\pi(n - 1)/2]}{n - 1} + \frac{\sin [\pi(n + 1)/2]}{n + 1} \right]$$

From this expression the following values are found: $a_2 = 1.06$, $a_3 = 0$, $a_4 = -0.21$, $a_5 = 0$, $a_6 = -0.09$, $a_7 = 0$, and $a_8 = -0.05$. Figure 6–16 shows this series expanded for the four nonzero harmonics 1, 2, 4, and 6. ■

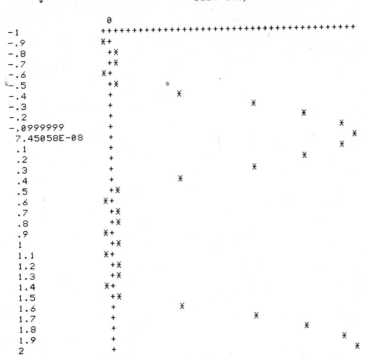

FIGURE 6–16. Computer plot of the Fourier series of Example 6-8 out to the first four nonzero harmonics.

Computer Aid: General.   Throughout this chapter the computer has been used as a general aid in the construction and plotting of Fourier series expansions. The plotting simply involves the use of plotting routines to display the shape of the series expansion expression. Also, the plotting of a series after the expansion amplitudes and/or phases have been modified gives a pictorial view of what effect the modification has had. This will be seen in Section 6-3. Another general use of the computer is for evaluation of the coefficients for the various harmonics from the equations that result from integration. This was done in Example 6-8 to determine the coefficients from the given expression. It took only a couple of moments to write a simple (BASIC) program to evaluate the a_n from the expression given.

Computer Aid: Integration.   In some cases a periodic function may be in the form of a graph for which an equation cannot be written even over a single period. In such a case it would be impossible to perform the integrations of Eqs. (6-4), (6-5), and (6-6) to find the Fourier series expansion. In these cases the integration is performed *numerically,* by the computer, using sets of data points $\{t, f(t)\}$, which describe the function over one period. Numerical integration consists of calculating the area of the curve over the specified range of the independent variable. Most computer software packages have integration programs both for functions and for data pairs. In the case of function integration the input is the function, the limits of integration, and the desired accuracy. In the case of data pair integration the input is simply the data pairs themselves. Some software packages are available which input such data pairs and output the desired Fourier series coefficients to whatever harmonic is requested. The following example assumes only that a data pair integration routine is available. In this case the calculation must be performed for each harmonic desired.

EXAMPLE 6-9

Find the Fourier series expansion for the function of Fig. 6–17 to the fifth harmonic.

SOLUTION

The function is periodic with a period of $T = 7$ s, but it is evident that no equation can be written for the function within one period. So numerical methods will be used. First, however, the function is seen to be *even* since $f(t) = f(-t)$. Then it will not be necessary to find the b_n since they will all be zero. Furthermore, it was pointed out earlier that for even functions the integration from $-T/2$ to $+T/2$ can be replaced by twice the integration from 0 to $T/2$. This means that the integrals are

$$a_0 = 2\left(\frac{1}{7}\right) \int_0^{3.5} f(t) \, dt$$

$$a_n = 2\left(\frac{2}{7}\right) \int_0^{3.5} f(t) \cos\left(\frac{n\pi t}{7}\right) \, dt$$

A series expansion to the fourth harmonic means that $n = 1, 2, 3,$ and 4. So there are actually five integrals to be evaluated. Let us take data points every 0.2 s, except for the last point at 3.5 s. Then the values of $f(t)$ are read off the graph every 0.2 s and tabulated. Next, these values of $f(t)$ are multiplied by $\cos(n\pi t/7)$ for $n = 1, 2, 3,$ and 4 and tabulated. The result is five sets of data pairs versus time. Each of these is to be integrated. A simple computer program was written which constructed these products from the input of t and $f(t)$. The

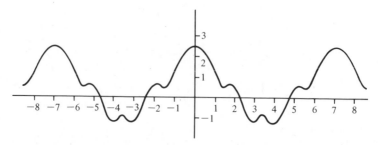

FIGURE 6–17. Signal for Example 6-9.

results were shown in Table 6–1. The integration program is used with each set of data points and the result gives the coefficients as follows:

$$a_0 = 0.486$$

$$a_1 = 1.645$$

$$a_2 = 0.131$$

$$a_3 = 0.286$$

$$a_4 = 0.184$$

$$a_5 = -0.271$$

TABLE 6–1

t	$f(t)$	p_1	p_2	p_3	p_4
0	2.5	2.5	2.5	2.5	2.5
.2	2.4	2.36143	2.24696	2.06028	1.80737
.4	2.35	2.20015	1.76972	1.11359	.315451
.6	2.1	1.80274	.995126	.0942128	−1.15688
.8	1.7	1.28022	.228199	−.936522	−1.63874
1	1.1	.68584	−.244771	−.991064	−.991067
1.2	.6	.284322	−.330537	−.597585	−.235817
1.4	.45	.139058	−.364057	−.364059	.139056
1.6	.5	.0671172	−.481981	−.196514	.429223
1.8	.6	−.0269179	−.597585	.0805374	.590358
2	.5	−.11126	−.450485	.311743	.311747
2.2	.2	−.0786047	−.138213	.187247	−8.97152E-03
2.4	−.2	.110179	.0786057	−.196786	.138211
2.6	−.8	.552849	.0358954	−.60246	.796779
2.8	−1.15	.930368	−.355364	−.355377	.930376
3	−1.2	1.08116	−.748183	.267016	.267037
3.2	−1	.963962	−.858446	.691056	−.47386
3.4	−.9	.896377	−.885536	.867564	−.842608
3.5	−.9	.9	−.9	.9	−.9

Results of the Integration

Harmonic number 0 the integral is 1.7   the coef is   .971429/2
Harmonic number 1 the integral is 2.87798   the coef is 1.64456
Harmonic number 2 the integral is   .229147 the coef is   .130941
Harmonic number 3 the integral is   .500513 the coef is   .286007
Harmonic number 4 the integral is   .322665 the coef is   .18438
Harmonic number 5 the integral is − .474034 the coef is − .270877
Harmonic number 6 the integral is − .0358001 the coef is − .0204572
Harmonic number 7 the integral is − .0206784 the coef is − .0118162
Harmonic number 8 the integral is   .0202843 the coef is   .011591

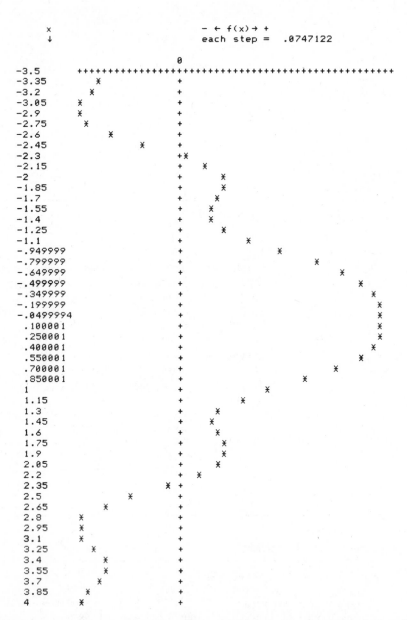

FIGURE 6–18. Computer plot of the first five harmonics of the Fourier series of the signal in Fig. 6–17.

This function is plotted in Fig. 6–18. You can see that the series expansion does indeed form the same function. ▪

APPLICATIONS OF FOURIER SERIES

One of the most important applications of Fourier series expansions of periodic functions is to display the frequency dependence of the signal via the spectrum. Another important application is to enable network analysis to be performed in those cases when the signal is input to some network and an analytic expression for the signal is required. In this section examples are given of these applications.

6-3.1 Frequency Response Effects

A common specification of electronic networks is the *frequency response*. This refers to the way in which the network responds to input signals of varying frequency. In some cases the network is purposely designed to pass or block certain frequencies and the frequency response is part of the desired design specification. These are *filter* circuits. In other cases one may have a network whose function is to amplify an input signal. Thus the output is supposed to be the same as the input but increased in amplitude. In most cases it is found that the amplification, or *gain,* of the network is not constant with respect to frequency. For example, the gain may be 10 at 1 kHz but only 6 at 10 kHz. Now, this would not matter much for pure sinusoidal signals, except that the output would not be as great for the higher-frequency inputs. However, a nonsinusoidal signal has been shown to be composed of a spectrum of many frequencies. If such a signal is input to the amplifier, the output signal will be *distorted* because not all frequency components received the same amplification. In this case the frequency response is an undesirable part of the design.

The following two examples show how the Fourier series expansion of signals can be used to study the effects of filters and amplifier frequency response. To define the response in an analytical fashion, use is made of the *transfer function* of the network. In general, the transfer function tells how the output of a network is determined by the input. For the present studies the transfer function will be the voltage gain of the network. The voltage gain is commonly expressed in two ways.

The *absolute* voltage gain simply reports the transfer function as the ratio of the output voltage magnitude to the input voltage magnitude:

$$G_v = \frac{|v_{out}|}{|v_{in}|} \tag{6-15}$$

Another common expression of gain is in terms of the power gain in decibels. Since the power delivered into a resistive load varies as the square of the voltage, this type of gain has an extra factor of 2:

$$A = 20 \log\left(\frac{|v_{out}|}{|v_{in}|}\right) \tag{6-16}$$

Thus a voltage gain of 100, meaning that the amplitude of the output voltage is 100 times the amplitude of the input voltage, would be equivalent to a power gain of $A = 20 \log (100) = 40$ dB. Note that if the G_v is less than 1, A will be negative and also that if $G_v = 1$, $A = 20 \log (1) = 0$ dB. Often, as noted above, the gain will be a function of frequency, and it is this effect that we will study using Fourier series expansion.

Phase Distortion. It should be noted that the phase of a signal is often also affected by the transfer function of a network as a function of frequency. You can see from the cosine representation of the Fourier series that if the phase is changed differently for different frequencies, the signal shape will be distorted.

EXAMPLE 6-10

The network represented in Fig. 6–19 is a high-pass filter. This means that for the lower frequencies the gain is less than 1 (attenuation) and at the high frequencies the gain is 1 (no effect). The diagram shows the gain of the network in decibels as a function of angular frequency. It is assumed that the phase is not affected. If the input voltage is the periodic signal of Fig. 6–4, find the shape of the output signal. Find the line spectrum and compare with that of the unfiltered signal.

SOLUTION

The Fourier series of this function has been found in both Examples 6-2 and 6-3. It is easier to use the cosine representation, so the series given in Example 6-3 will be used. The solution is simply to find the new amplitude of each harmonic by multiplying the original amplitude by the gain found from the curve of Fig. 6–19. This has been done and tabulated in Table 6–2. Note that the low-frequency coefficients have been strongly attenuated. The output signal is plotted using the Fourier series which results after the effects on the coefficients have been computed. This is shown in Fig. 6–20a. Note that the low-frequency variations are gone, leaving only abrupt level changes corresponding to the rapid changes of the input signal. In effect, the output now provides pulses for every transition of the input signal. The line spectrum of Fig. 6–20b shows clearly that the low-frequency components have been reduced to a small fraction of their original form. ∎

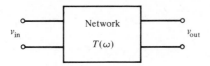

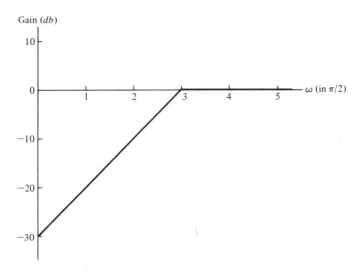

FIGURE 6–19. Network and transfer function for Example 6-10.

TABLE 6–2

n	c_n (input)	Gain (dB)	Gain (absolute)	c_n (output)
0	3.75	−30	0.032	0.12
1	2.25	−20	0.1	0.23
2	1.59	−10	0.32	0.51
3	0.75	0	1	0.75
4	0.0	0	1	0.0
5	0.45	0	1	0.45
6	0.53	0	1	0.53
7	0.32	0	1	0.32
8	0	0	1	0
9	0.25	0	1	0.25
10	0.32	0	1	0.32
11	0.21	0	1	0.21
12	0	0	1	0
13	0.17	0	1	0.17
14	0.23	0	1	0.23
15	0.15	0	1	0.15
16	0	0	1	0

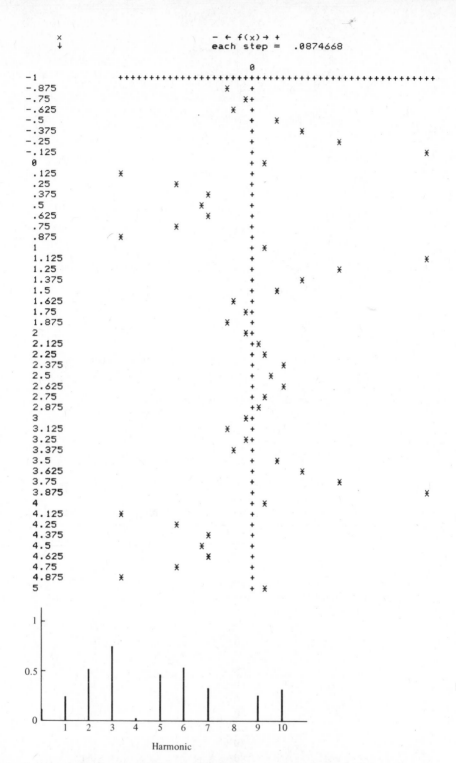

FIGURE 6-20. Above: the output signal Fourier series is plotted from Example 6-10. The amplitude line spectrum of this output is shown below.

FOURIER ANALYSIS

EXAMPLE 6-11

The network represented in Fig. 6–21 is an amplifier. The gain plot in decibels versus frequency shows that the gain is a constant 20 dB (10 absolute) from 0 to 2 rad/s and then drops off at 10 dB per rad/s. The gain has fallen to 0 dB at 5 rad/s. It is assumed the phase is not affected. Assume that the signal of Fig. 6–17 is input to this amplifier. Find the shape of the output signal.

SOLUTION

The Fourier series coefficients were found in Example 6-9 for the signal in Fig. 6–17. It is now a matter of multiplying each coefficient by the gain at that frequency found from the graph in Fig. 6–21. The result is

$$a_0 = 0.486 \times 10 = 4.86$$

$$a_1 = 1.645 \times 10 = 16.45$$

$$a_2 = 0.131 \times 10 = 1.31$$

$$a_3 = 0.286 \times 10 = 2.86$$

$$a_4 = 0.184 \times 3.16 = 0.58$$

$$a_5 = -0.271 \times 1 = -0.27$$

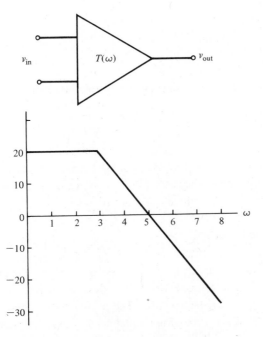

FIGURE 6–21. Amplifier and transfer function for Example 6-11.

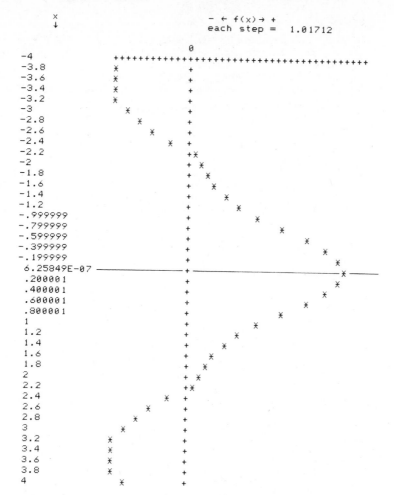

```
         x                       - ← f(x) → +
         ↓                       each step =   1.01712

                                      0
    -4        +++++++++++++++++++++++++++++++++++++++++++++++++++
    -3.8           *                 +
    -3.6           *                 +
    -3.4           *                 +
    -3.2           *                 +
    -3              *                +
    -2.8              *              +
    -2.6               *             +
    -2.4                *            +
    -2.2                             +*
    -2                               + *
    -1.8                             +  *
    -1.6                             +   *
    -1.4                             +     *
    -1.2                             +       *
    -.999999                         +          *
    -.799999                         +            *
    -.599999                         +              *
    -.399999                         +                *
    -.199999                         +                  *
   6.25849E-07 ─────────────────────────+──────────────────────────*───
    .200001                             +                           *
    .400001                             +                        *
    .600001                             +                     *
    .800001                             +                 *
   1                                    +             *
   1.2                                  +          *
   1.4                                  +        *
   1.6                                  +     *
   1.8                                  +   *
   2                                    + *
   2.2                                  +*
   2.4                           *      +
   2.6                       *          +
   2.8                    *             +
   3                 *                  +
   3.2           *                      +
   3.4           *                      +
   3.6           *                      +
   3.8           *                      +
   4               *                    +
```

FIGURE 6–22. Output of Example 6-11, showing the distortion of the input signal given by Fig. 6–17.

Figure 6–22 shows a plot of the Fourier series expansion using the coefficients found above. You can see that the original signal has been distorted by the drop in gain with frequency. In particular, as noted from the gain versus frequency, the higher-frequency variations of the signal have been lost. ■

6-3.2 Network Analysis

Application of Fourier series to network analysis is based on the ability to express a periodic function by an analytical expression which can then be used in the subsequent analysis. In essence, network analysis is finding the transfer function introduced in the preceding section. The usefulness of Fourier series expansions in network analysis is based in part on a very important principle of electrical networks.

Superposition. In Fig. 6–23 a series of sources have been applied to the input of a linear network of the type which has been the subject of this text. The input is

$$v_{in} = v_{1in} + v_{2in} + \cdots + v_{nin}$$

The response of the network is some output voltage, v_{out}. The principle of superposition states that the output can be expressed as a sum of terms, each of which is derived from the corresponding input. In other words, the network output can be solved for each input in the series, independently, and the total output will be the sum of the individual terms:

$$v_{out} = v_{1out} + v_{2out} + \cdots + v_{nout}$$

where

$$v_{1out} = \text{output for } v_{1in}$$

$$v_{2out} = \text{output for } v_{2in} \quad \text{etc.}$$

This means that the individual terms of a Fourier series can be considered as inputs to a network and the corresponding output found. The net output will then be the sum of these terms. The idea is like constructing the Fourier series expansion of the output voltage by finding out how the network affects each term of the Fourier series of the input voltage. The following example illustrates the basic procedure.

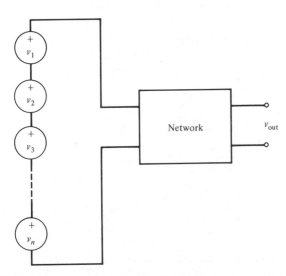

FIGURE 6–23. Illustration of how each Fourier term can be considered as a separate source for input to a network.

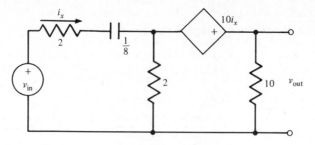

FIGURE 6–24. Network for Example 6-12.

EXAMPLE 6-12

The signal of Fig. 6–10 is input to the network shown in Fig. 6–24. Find and plot the output voltage. Assume no initial conditions.

SOLUTION

The first step in finding a solution to this problem will be to find the Laplace transform of the output voltage in terms of the input voltage, which is left unspecified, $V_{in}(s)$. The transformed network is shown in Fig. 6–25 with the loop currents identified for generalized mesh analysis. The required output voltage will be

$$V_{out}(s) = 10I_2(s)$$

The matrix equation is

$$\begin{pmatrix} 4 + \dfrac{8}{s} & -2 \\ -2 & 12 \end{pmatrix} \begin{pmatrix} I_1 \\ I_2 \end{pmatrix} = \begin{pmatrix} V_{in} \\ 10I_x \end{pmatrix}$$

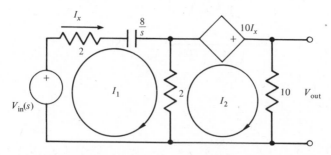

FIGURE 6–25. Network of Fig. 6–24, transformed and prepared for mesh analysis.

From the network it is clear that $I_x = I_1$, so the matrix equation becomes

$$\begin{pmatrix} 4 + \dfrac{8}{s} & -2 \\ -12 & 12 \end{pmatrix} \begin{pmatrix} I_1 \\ I_2 \end{pmatrix} = \begin{pmatrix} V_{in} \\ 0 \end{pmatrix}$$

Cramer's rule can be used to solve for I_2 and V_{out}:

$$V_{out}(s) = \frac{5sV_{in}(s)}{s + 4}$$

Now, to find the output voltage, each term of the input Fourier series can be substituted into this equation and the corresponding output solved. It may be easier, however, to note first that because of the capacitor the output cannot have a dc term. Thus the dc term of the input, $a_0 = 0.5$, will contribute no output term, $V_{0out}(s) = 0$. The series, as found in Example 6-6, contains only cosine terms and, in general,

$$V_{nin}(t) = a_n \cos (n\pi t)$$

where it was shown that $\omega_0 = \pi$. This means that the general Laplace input voltage will be

$$V_{nin}(s) = \frac{sa_n}{s^2 + (n\pi)^2}$$

The output voltage, for the nth term, can now be written:

$$V_{nout}(s) = \frac{5a_n s^2}{(s + 4)[s^2 + (n\pi)^2]}$$

In finding the partial fraction expansion it must be assumed that both sine and cosine terms will be present in the final solution, so the result is

$$V_{nout}(s) = \frac{K_1}{s + 4} + \frac{n\pi B_n}{s^2 + (n\pi)^2} + \frac{sA_n}{s^2 + (n\pi)^2}$$

where A_n is the coefficient of the cosine term in the nth harmonic and B_n the coefficient of the sine term. The method of matching is now used to find equations for these coefficients.

$$A_n = \frac{5(n\pi)^2 a_n}{16 + (n\pi)^2} \qquad B_n = \frac{-20n\pi a_n}{16 + (n\pi)^2}$$

The fact that the original series had no sine terms but the output does show that a phase shift has occurred.

The Fourier series of the output can now be constructed by evaluating the coefficients above for $n = 1$, 3, and 5 (for evaluation out to the fifth harmonic). The values of a_n are found from Example 6-6.

$$n = 1 \qquad A_1 = 2.32 \qquad B_1 = -2.96$$

$$n = 3 \qquad A_3 = 0.57 \qquad B_3 = -0.24$$

$$n = 5 \qquad A_5 = 0.23 \qquad B_5 = -0.06$$

So now the time solution, ignoring the transient, can be expressed as a Fourier series out to the fifth harmonic:

$$v_{out}(t) = 2.32 \cos (\pi t) - 2.95 \sin (\pi t) + 0.57 \cos (3\pi t) - 0.24 \sin (3\pi t)$$
$$+ 0.23 \cos (5\pi t) - 0.06 \sin (5\pi t)$$

This function is plotted in Fig. 6–26. The distortion due to the variation in gain and phase shift with frequency is apparent. ∎

6-4

FOURIER TRANSFORMS

The purpose of this section is to give a definition of the Fourier transform. It would be far beyond the scope of this text to present a practical treatment of Fourier transform construction, application, and interpretation. There is value, however, to introducing this important signal descriptor and showing its relationship to Fourier series and Laplace transforms.

6-4.1 Introduction

The Fourier series expansion of a periodic signal provided a description of that signal which shows its frequency spectrum. The spectrum was found to be in harmonics of the fundamental frequency, ω_0. The series expansion also allowed an equation to be written which described the signal for all time. This means that the signal could be reconstructed by considering combinations of oscillations at frequencies, ω_0, $2\omega_0$, $3\omega_0$, $4\omega_0$, ... The fundamental frequency was determined by the period of the signal, T, $\omega_0 = 2\pi/T$. Fourier transforms results from the desire to make such a frequency spectrum representation of a *non-periodic* signal. This means a signal that consists of a single event such as a pulse or a "burst" of voltage variation over a finite time span. To see how such a description is constructed, it will first be necessary to consider yet another representation of the Fourier series.

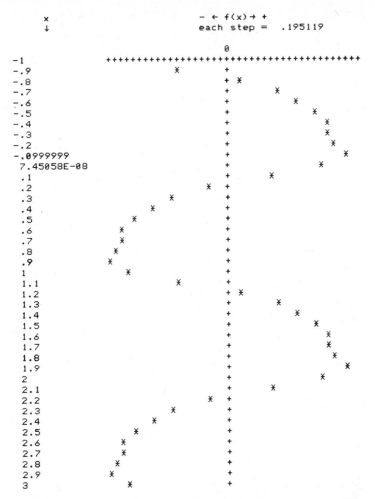

FIGURE 6—26. Output signal Fourier series plotted by computer for Example 6-12.

Complex Fourier Series.   You will recall that it is possible to represent sine and cosine functions by exponentials which have a complex exponent. This is defined by the Euler relations,

$$\sin (\omega t) = \frac{e^{j\omega t} - e^{-j\omega t}}{2j} \qquad (6\text{-}17)$$

$$\cos (\omega t) = \frac{e^{j\omega t} + e^{-j\omega t}}{2} \qquad (6\text{-}18)$$

and the reverse relation,

$$e^{\pm j\omega t} = \cos (\omega t) \pm j \sin (\omega t) \qquad (6\text{-}19)$$

Now the relations of Eqs. (6-17) and (6-18) will be substituted into the definition of the Fourier series as given by Eq. (6-3), giving

$$f(t) = a_0 + \sum_{n=1}^{\infty} a_n \frac{e^{jn\omega_0 t} + e^{-jn\omega_0 t}}{2} + b_n \frac{e^{jn\omega_0 t} - e^{-jn\omega_0 t}}{2j}$$

This expression can be regrouped to form the equation

$$f(t) = a_0 + \sum_{n=1}^{\infty} \frac{a_n - jb_n}{2} e^{jn\omega_0 t} + \sum_{n=1}^{\infty} \frac{a_n + jb_n}{2} e^{-jn\omega_0 t}$$

From an examination of the equations for a_n and b_n given by Eqs. (6-5) and (6-6), it is clear that if n is replaced by $-n$, then $a_{-n} = a_n$ since $\cos(-x) = \cos(x)$ and also $b_{-n} = -b_n$ since $\sin(-x) = -\sin(x)$. So in the second sum above, n is replaced by $-n$. The series now runs from -1 to $-\infty$. When this is done the two sums and the $n = 0$ term will combine to run from $-\infty$ to $+\infty$. The *complex form* of the Fourier series expansion can be written

$$f(t) = \sum_{n=-\infty}^{\infty} C_n e^{jn\omega_0 t} \tag{6-20}$$

where

$$C_0 = a_0 \tag{6-21}$$

$$C_n = \frac{a_n - jb_n}{2}$$

Actually, by using the Euler relations you can show that the coefficients given by Eq. (6-21) can be found from a single integral given by

$$C_n = \frac{1}{T} \int_0^T f(t) e^{-jn\omega_0 t}\, dt \tag{6-22}$$

This equation even includes the $n = 0$ term. There is a certain satisfaction to seeing how the introduction of this new notation results in such compact relations as Eqs. (6-20) and (6-21).

Of course, the coefficients that result from Eq. (6-22) will be complex numbers. The actual line spectrum amplitudes, which are the c_n of the cosine representation, are found from the relations between this complex representation of the Fourier series and the cosine representation:

$$c_0 = C_0 \tag{6-23}$$

$$c_n = 2\,|C_n| = 2[(\mathrm{Re}\ (C_n)^2 + (\mathrm{Im}\ (C_n)^2]^{1/2}$$

$$\phi_n = -\tan^{-1}\left[\frac{\mathrm{Im}\ (C_n)}{\mathrm{Re}\ (C_n)}\right] \tag{6-24}$$

where Re and Im refer to the real part and imaginary part of C_n. For completeness it should be clear that

$$a_n = 2\ \mathrm{Re}\ (C_n) \qquad \text{and} \qquad b_n = -2\ \mathrm{Im}\ (C_n)$$

EXAMPLE 6-13

The periodic signal of Fig. 6–27 is typical of relaxation type of oscillators which depend on the discharging of a capacitor. The signal from 0 to 2 s is described by the equation

$$f(t) = 10e^{-t}$$

Find the complex Fourier series. Use the cosine representation to plot the function to nine harmonics.

SOLUTION

Evidently, the period is $T = 2$ s, so the fundamental frequency is $\omega_0 = \pi$ rad/s. The coefficients of the complex series are found from Eq. (6-22):

$$C_n = \frac{1}{2}\int_0^2 10e^{-t}e^{-jn\pi t}\ dt$$

$$= -\frac{5}{1 + jn\pi}\,e^{-(1+jn\pi)t}\Bigg]_0^2$$

$$= -\frac{5}{1 + jn\pi}\,(e^{-2}e^{-2jn\pi} - 1)$$

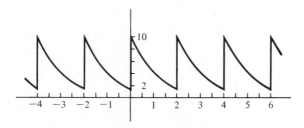

FIGURE 6–27. Signal for Example 6-13.

but $e^{-2jn\pi} = \cos(2n\pi) - j\sin(2n\pi) = 1$ and $e^{-2} = 0.135$:

$$C_n = \frac{4.32}{1 + jn\pi} = \frac{4.32}{1 + jn\pi} \frac{1 - jn\pi}{1 - jn\pi}$$

Thus the complex Fourier coefficient is

$$C_n = \frac{4.32}{1 + (n\pi)^2} (1 - jn\pi)$$

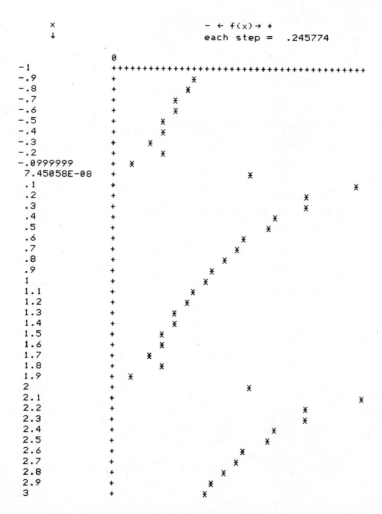

FIGURE 6—28 (a). Computer plot of the Fourier series representation of the signal of Fig. 6—27.

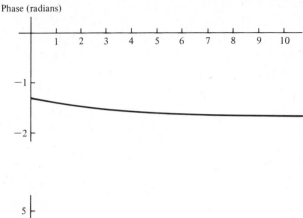

Phase (radians)

Harmonic

FIGURE 6–28 (b). Line spectrum of amplitude and phase of the signal shown in Fig. 6–27.

The cosine representation coefficients are found from Eqs. (6-23) and (6-24) as

$$c_0 = C_0 = 4.32$$

$$c_n = 2\,|C_n| = \frac{8.64}{[1 + (n\pi)^2]^{1/2}}$$

$$\phi_n = \tan^{-1}(-n\pi)$$

The cosine representation is plotted in Fig. 6–28 together with the line spectrum. ∎

The complex coefficient found in this example has values for n from $-\infty$ to $+\infty$. You must remember that the negative harmonics are simply a mathematical construction to allow the complex series to be written in an abbreviated form. The actual physically identifiable quantities are the cosine representations which are valid only for positive values of n.

6-4.2 Fourier Transform

The concept of the Fourier transform can be developed from the complex Fourier series presented above. First, let us write the definition of the complex Fourier coefficient, Eq. (6-22), with limits extending from $-T/2$ to $+T/2$, which extends over one full period and is therefore a valid limit range on the integral. An angular frequency variable is defined by

$$\omega = n\omega_0 \qquad (6\text{-}25)$$

Note that this variable takes on all values of harmonic frequency as n varies from $-\infty$ to $+\infty$ in increments of ω_0. Now, if the period is allowed to go to infinity, $T \to \infty$, notice that the angular frequency increment will go to zero, $\omega_0 \to 0$; in other words, it becomes *infinitesimal*. Thus, in this limit, the value of the angular frequency defined by Eq. (6-25) will sweep out *all* values as n sweeps from $-\infty$ to $+\infty$; in other words, it becomes a *continuous variable*. In this case it becomes unnecessary to speak of a "harmonic" of the fundamental frequency but rather of just the angular frequency as it varies smoothly from $-\infty$ to $+\infty$. The equation for harmonic coefficient now becomes an equation for angular frequency coefficient:

$$F(\omega) = \lim_{T \to \infty} (TC_n) = \int_{-\infty}^{\infty} f(t)e^{-j\omega t}\, dt \qquad (6\text{-}26)$$

Instead of a line spectrum we now have a continuous spectrum of amplitude versus frequency. There is no requirement that the function $f(t)$ in Eq. (6-26) be periodic in time. Any time function can be transformed which satisfies the conditions of a finite number of discontinuities, finite discontinuity amplitude, and the integral relation of finite absolute area under the curve, now extended from $-\infty$ to ∞.

Equation (6-26) can now be used to find the frequency spectrum of any signal, periodic or nonperiodic. In the case of Fourier transforms the idea of obtaining the frequency spectrum becomes more important than that of obtaining an analytical expression for the time function, but both concepts are still present.

The *inverse* relation to Eq. (6-26) will be an equation that shows how to reconstruct the time function from the Fourier transform. This is provided by considering the change in the series expression of Eq. (6-20) as $\omega_0 \to 0$. Prior to passing the limit of $T \to 0$, C_n and $F(n\omega_0)$ are related by the expression

$$C_n = \frac{F(n\omega_0)}{T} = \frac{\omega_0 F(n\omega_0)}{2\pi}$$

This is now substituted into Eq. (6-20) and the limit is passed as $T \to \infty$ and therefore $n\omega_0 \to \omega$, a continuous variable, and $\omega_0 \to d\omega$, an infinitesimal increment of angular frequency. This process gives

$$f(t) = \lim_{T\to\infty} \sum_{n=-\infty}^{\infty} \frac{\omega_0}{2\pi} F(n\omega_0)e^{jn\omega_0 t}$$

Since the process has become infinitesimal, the summation becomes integration. The final expression for $f(t)$ becomes

$$f(t) = \frac{1}{2\pi} \int_{-\infty}^{+\infty} F(\omega)e^{j\omega t}\, d\omega \qquad (6\text{-}27)$$

EXAMPLE 6-14

Find the Fourier transform and frequency spectrum of the single pulse of Fig. 6–29.

SOLUTION

The solution is provided by integration of the $f(t)$ of the pulse according to the equation for $F(\omega)$ given by Eq. (6-26). The function to be integrated is given by

$$f(t) = \begin{cases} 10 & \text{for } -2 < t < +2 \\ 0 & \text{everywhere else} \end{cases}$$

Then Eq. (6-26) becomes

$$F(\omega) = \int_{-a}^{a} 10e^{-j\omega t}\, dt = -\frac{10}{j\omega} e^{-j\omega t}\Bigg]_{-a}^{a}$$

$$= -\frac{10}{j\omega}(e^{-j\omega a} - e^{j\omega a}) = \frac{20}{\omega}\frac{e^{j\omega a} - e^{-j\omega a}}{2j}$$

$$= 20\,\frac{\sin(\omega a)}{\omega}$$

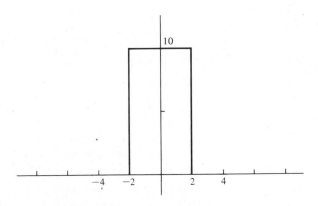

FIGURE 6–29. Signal for Example 6-14.

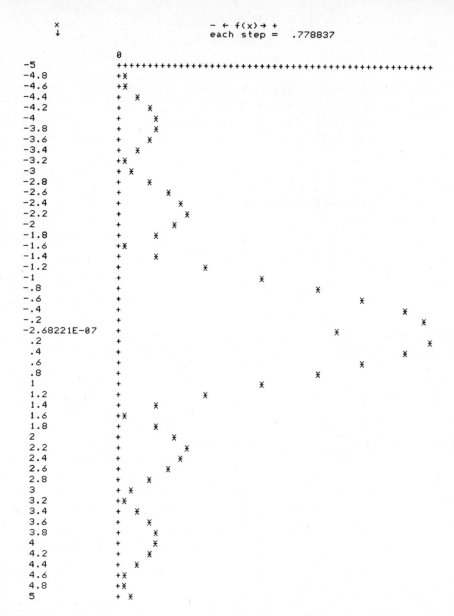

FIGURE 6–30. Continuous spectra from the Fourier transform of the signal in Fig. 6–29.

This equation gives the continuous amplitudes of the angular frequency that makes up the time signal. It is to be compared to the line spectrum amplitudes of Fourier series. A plot of the spectrum uses the absolute value of $F(\omega)$ so that comparisons of different frequency components will be clearer. This function will be plotted for all values of ω from $-\infty$ to $+\infty$, but you should remember that the negative frequencies came about as a result of mathematical convenience and have no particular physical significance. In fact, it is the sum of the positive and negative frequency amplitudes which are significant. Figure 6–30 shows the spectrum for this pulse. ■

Relation to Laplace Transforms. For most functions there is a simple relationship between the Fourier transform and the Laplace transform. A comparison between Eq. (4-3) for the Laplace transform and Eq. (6-26) for the Fourier transform shows the difference. In particular, you will recall that the Laplace transform is defined only for functions in the range $t > 0$. In fact, all Laplace functions can be considered to be multiplied by the step function $f(t)u(t)$ prior to transformation. This is not true for the Fourier transform. Also, there seems to be a change of variable since the Fourier transform involves $j\omega$ instead of s. These differences can be used to relate the two transforms. In general, if s is replaced by $j\omega$ in a Laplace transform, the Fourier transform results. For example, for the time function $f(t) = e^{-at} \cos(\omega_0 t)$, the Laplace transform is

$$F(s) = \frac{s + a}{(s + a)^2 + (\omega_0)^2}$$

The Fourier transform can be found by replacing s by $j\omega$:

$$F(\omega) = \frac{j\omega + a}{(j\omega + a)^2 + (\omega_0)^2}$$

The reverse is also true, so the Laplace transform can be found by replacing $j\omega$ by s in a Fourier transform. For example, the function $f(t) = u(t)e^{-at}$ has the Fourier transform

$$F(\omega) = \frac{1}{j\omega + a}$$

Now, by substitution of s for $j\omega$,

$$F(s) = \frac{1}{s + a}$$

which you will recognize as the Fourier transform of the exponential. Care must be taken in such substitutions that the conditions for existence of the Fourier transform are not violated and that the function is defined for $t > 0$ in the Laplace transform.

Discrete Fourier Transform and the Computer. As in all areas of technology, the computer is used extensively in Fourier transform analysis. In this case, however, there is a particular application, other than the standard plotting and integrating, which deserves mention in this chapter. This has to do with the desire to find the Fourier transform (i.e., the frequency spectrum) of discrete data points from some real-world variable, which shows the burst or single-event type of behavior. A typical example is that of the voltage oscillations which result from a word spoken into a microphone.

In these cases a computer is used to take periodic samples of the voltage in time during the existence of the signal. These samples are to be used to find the frequency spectrum of the signal. As might be expected, such a set of data can provide only a discrete number of frequency components. The integrals of Eqs. (6-26) and (6-27) revert to sums over the discrete data, as evaluated numerically by the computer. The new expressions become:

Discrete Fourier Transform. *Given a discrete set of N data points, f_n, n = 0, 1, 2, . . ., N − 1, and each separated in time by interval, Δt, the discrete Fourier transform (DFT) is defined by*

$$F_m = \sum_{n=0}^{N-1} f_n e^{-jmn\Delta\omega\Delta t} \qquad (6\text{-}28)$$

where

$m = 0, 1, 2, . . ., N − 1$, *so there will be N terms in the transform.*

$\Delta\omega = 2\pi/(N\,\Delta t) =$ *frequency interval*

The inverse relation is

$$f_n = \frac{1}{N} \sum_{m=0}^{N-1} F_m e^{jnm\Delta\omega\Delta t} \qquad (6\text{-}29)$$

These equations are just the numerical integration of the Fourier transform equations expressed in a form compatible with computer usage. It is important to note that the result will be a discrete spectrum since if one has a finite number of data points, there can be but finite knowledge of possible frequency content. Numerous computer algorithms have been developed to evaluate the summation of Eq. (6-28) to determine the frequency analysis of a signal. One of the most efficient is called the fast Fourier transform (FFT), which stipulates that the number of samples N of the time signal must be modulo 2. This means that $N = 2, 4, 8, 16, 32, ...$

The Fourier coefficients found from Eq. (6-28) will, in general, be complex numbers. The magnitude of the complex pair for the discrete frequencies represents the spectrum of the signal and the phase tells how each frequency component is related in a time reconstruction. The full set of N data points will give only half as many unique Fourier coefficients and resulting spectral information.

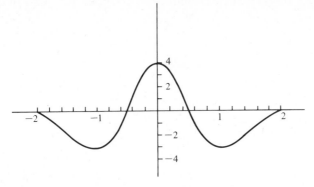

FIGURE 6–31. Signal for Example 6-15.

This is because the Fourier coefficients, being complex numbers, require two calculations per data point. This result can be traced back to the development of the complex Fourier series when the series was extended from ω going from 0 to ∞ to variation from $-\infty$ to $+\infty$.

EXAMPLE 6-15

Find the DFT of the pulse shown in Fig. 6–31. Assume that data are taken every 0.2 s.

SOLUTION

Since data are taken every 0.2 s, $\Delta t = 0.2$ and also there will be 21 data points, starting at -2 and ending at $+2$. This means that the increment of angular frequency is $\Delta \omega = 2\pi/(N\,\Delta t) = 1.5$ rad/s. The 21 data points are shown in Table 6–3 as read from the graph. Now the transform will also have 21 terms calculated from the series found from Eq. (6-28):

$$F_m = f_0 + f_1 e^{-0.3mj} + f_2 e^{-0.6mj} + \cdots + f_{20} e^{-6mj} + f_{21} e^{-6.3mj}$$

TABLE 6-3

t	$f(t)$	t	$f(t)$
-2	0	0.2	3
-1.8	-0.4	0.4	1
-1.6	-1.25	0.6	-1.25
-1.4	-2.1	0.8	-2.5
-1.2	-2.75	1	-3
-1	-3	1.2	-2.75
-0.8	-2.5	1.4	-2.1
-0.6	-1.25	1.6	-1.25
-0.4	1	1.8	-0.4
-0.2	3	2	0
0	4		

Using the Euler relations, this equation can be expressed as two equations, one for the real part, in terms of cosine terms, and one for the imaginary part, expressed as sine terms.

$$\text{Re } (F_m) = f_0 + f_1 \cos (0.3m) + f_2 \cos (0.6m) + \cdots + f_{21} \cos (6.3t)$$

$$\text{Im } (F_m) = -f_1 \sin (0.3t) - f_2 \sin (0.6t) - \cdots - f_{21} \sin (6.3t)$$

It is a simple matter to write a computer program to calculate these terms for $m = 0, 1, 2, \ldots, 11$ using the data given in Table 6–3. Note that it is not necessary to perform the calculation for the remaining values of m because this will simply repeat the first set. Table 6–4 shows the results. The table includes the magnitude so that a spectrum can be constructed, as shown in Fig. 6–32. Note that there is a significant peak at $m = 2$. This would correspond to a signal of angular frequency, $\omega = m \Delta\omega = (2)(1.5 \text{ rad/s}) = 3 \text{ rad/s}$. This in turn represents a sinusoidal oscillation with a period of $T = 2\pi/\omega = 2.1$ s. You can see from the signal of Fig. 6-31 that this represents the period of the basic signal oscillation and we would thus expect a spectral peak at this frequency. ∎

TABLE 6–4

m	Re (F_m)	Im (F_m)	F_m
0	− 15.3	0	15.3
1	− 13.1479	− 1.98175	13.2964
2	24.985	7.707	26.1467
3	− 5.794	− 2.7903	6.43088
4	1.69335	1.15453	2.04949
5	− .105995	− .09835	.144595
6	− .181476	− .22758	.291077
7	.224986	.389715	.449997
8	− 9.96547E-03	− .0254023	.0272872
9	− .0245375	− .107495	.11026
10	.0104667	.139775	.140166
11	.010467	− .139787	.140178
12	− .0245377	.107477	.110243
13	− 9.9753E-03	.0254017	.0272901
14	.225023	− .389704	.450005
15	− .181507	.227573	.291091
16	− .105966	.0983367	.144565
17	1.69338	− 1.15444	2.04946
18	− 5.79389	2.78987	6.43059
19	24.9852	− 7.70557	26.1464
20	− 13.1487	1.98116	13.2971

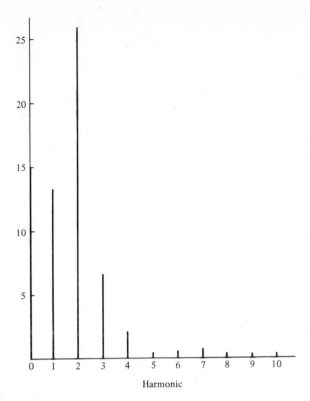

FIGURE 6–32. Spectrum of the discrete Fourier transform of the signal in Fig. 6–31.

SUMMARY

This chapter presented the essential features of Fourier series analysis of electrical signals, as well as an introduction to Fourier transforms and discrete Fourier expansions. To do this a number of specific issues were treated, including:

1. A Fourier series was shown to be a series expansion in sine and/or cosine trigonometric functions. The series is valid for periodic functions.
2. The expansion is in integer multiples, called *harmonics* of fundamental oscillation frequency of the signal.
3. Integral relations were given by which the coefficients of the harmonics can be determined from information about the value of the function over one period.
4. The spectrum and power spectrum of a signal Fourier series gives information about the relative degree of harmonic contribution to a signal.
5. The level symmetry of a signal allows easy, geometric determination of the dc level of some signals.

6. Functions that have *even* time symmetry, $f(t) = f(t + T)$, do not have any sine function terms in the series expansion $b_n = 0$.
7. Functions that have *odd* time symmetry, $f(t) = -f(t + T)$, do not have any cosine function terms in the series expansion $a_n = 0$.
8. Fourier transforms were shown to develop from Fourier series in the limit that the period goes to infinity. The Fourier transform and Laplace transform were shown to be related by $s = j\omega$.
9. The discrete Fourier transform was shown to be an expression of the Fourier transform constructed from a discrete number of data points.

PROBLEMS

6–1   Given the power series for log $(1 + x)$,

$$\log (1 + x) = x - \frac{x^2}{2} + \frac{x^3}{3} - \frac{x^4}{4} + \frac{x^5}{5} - \cdots (x^2 < 1)$$

(a) Find the number of terms so that the next term error at $x = 0.5$ is less than 1% compared to the actual value.
(b) Plot this series, to that number of terms, and log $(1 + x)$ from 0 to 1.

6–2   Derive Eq. (6-4) for the dc term in the Fourier series by integrating Eq. (6-3) from 0 to T.

6–3   Derive Eq. (6-5) for a_n by multiplying Eq. (6-3) on both sides by cos $(n\omega_0 t)$ and integrating from 0 to T.

6–4   Use the integral relation to find the Fourier series of the signal in Fig. 6–33. Plot the series to the first three nonzero harmonics. What is the maximum error in a period? Derive the cosine representation and plot the line spectrum to 10 harmonics. Plot the power spectrum. What percentage of the power in the first 10 harmonics is contained in the dc term and the first harmonic?

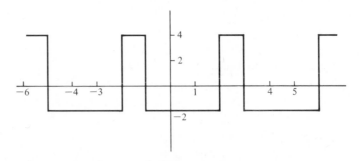

FIGURE 6–33.   Signal for Problem 6–4.

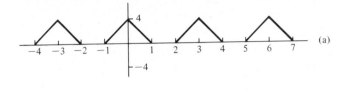

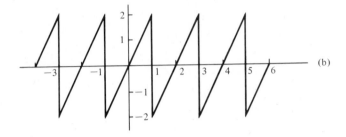

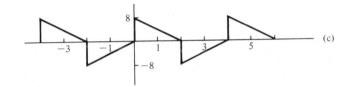

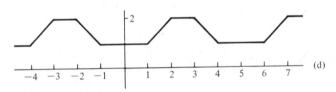

FIGURE 6–34. Signals for Problems 6–5, 6–6, 6–7, 6–8, 6–12, 6–14, and 6–15.

6–5   Find the period and any symmetry properties of the signals given in Fig. 6–34. State what the consequences of the symmetry properties of each is on the Fourier coefficients.

6–6   Use geometric principles to find the dc term, a_0, of each signal in Fig. 6–34.

6–7   Find the Fourier series to three nonzero harmonics for the signals given in Fig. 6–34a and b. Plot the line spectrum of each to 10 harmonics.

6–8   Find the Fourier series to three nonzero harmonics for the signals given in Fig. 6–34c and d. Plot the line spectrum of each to 10 harmonics.

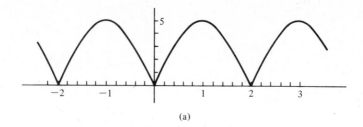

(a)

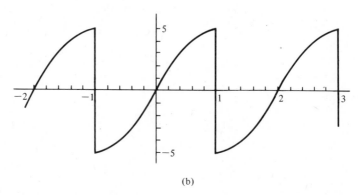

(b)

FIGURE 6–35. Signals for Problems 6–9, 6–10, 6–11, 6–13, and 6–16.

6–9   Find the Fourier series to three nonzero harmonics of the signal given in Fig. 6–35a. Find the cosine representation and plot the spectrum to 10 harmonics.

6–10   Find the Fourier series to three nonzero harmonics of the signal given in Fig. 6–35b. Find the line spectrum to 10 harmonics.

6–11   The signal of Fig. 6–35a is input to a low-pass filter with a transfer function as shown in Fig. 6–36. Assume that the phase is not affected. Find the cosine representation of the output signal Fourier series and plot. Compare the power spectrum of the input and output signal. How much effect did the filter have on the signal power? Explain.

6–12   An amplifier has a transfer function with respect to phase and gain as shown in Fig. 6–37. Determine the output signal Fourier series and plot if the input is the signal of Fig. 6–34d.

6–13   The network of Fig. 6–38 has an input as the signal of Fig. 6–35b. Calculate the cosine representation Fourier series of the output. Plot and compare to the input. Divide the input voltage magnitude by the output voltage magnitude at

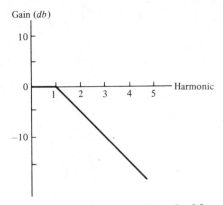

FIGURE 6–36. Transfer function for Problem 6–11.

each harmonic (up to 10). From this determine the gain in decibels at each harmonic and plot to determine the transfer function of the network. Find the phase transfer function.

6–14 Find the complex form of the Fourier series for the signal given in Fig. 6–34a. Plot the line spectrum.

6–15 Find the complex form of the Fourier series for the signal in Fig. 6–34d. Plot the line spectrum.

6–16 Find the complex form of the Fourier series of the signal in Fig. 6–35b, by integration. (*Hint:* Use the Euler relations to convert the trigonometric functions to exponentials prior to integration.)

6–17 Find the Fourier transform of the signal in Fig. 6–39. Plot the spectrum and compare with that found in Problem 6–14.

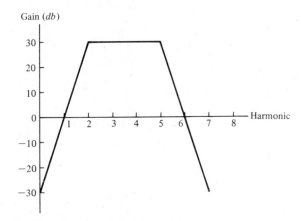

FIGURE 6–37. Transfer function for Problem 6–12.

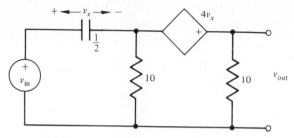

FIGURE 6–38.   Network for Problem 6–13.

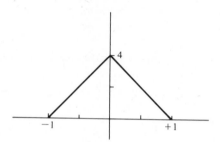

FIGURE 6–39.   Signal for Problems 6–17 and 6–19.

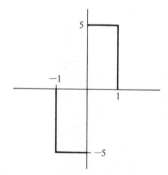

FIGURE 6–40.   Signal for Problem 6–18.

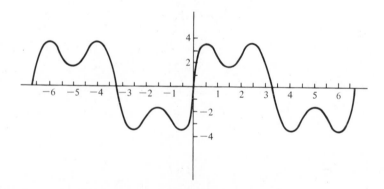

FIGURE 6–41.   Signal for Problem 6–20.

6–18   Find the Fourier transform of the signal in Fig. 6–40. Plot the spectrum of this signal.

6–19   Find the discrete Fourier transform of the signal in Fig. 6–39 using data every 0.2 s. Compare the spectrum with that found in Problem 6–17.

COMPUTER PROBLEMS

6–20   Find the Fourier series cosine representation of the signal shown in Fig. 6–41 out to 10 harmonics. Plot the series and compare with the actual signal. Plot the line spectrum of the signal.

6–21   Find the Fourier transform of the signal shown in Fig. 6–29. Convert the signal into many data points and use a computer integration routine. Convert the integrals to real and imaginary parts using the Euler relations. Plot the spectrum.

Solutions to Selected Odd-Numbered Problems

Chapter 1

1–1 $p(t) = 50 \cos (2t) \sin (2t)$, plotted in Fig. S–1

1–3 $p(t) = 50(1 - e^{-4t})e^{-4t}$
$\omega(t) = (50/4)(1 - e^{-4t}) - (50/8)(1 - e^{-8t})$
Plotted in Fig. S–2a and b.

1–5 $i(t) = 24t + 20$

1–7 $\omega = 2975.9$ J, $v = 50.9$ V

1–9 $\omega(0) = 24$ J, $\omega(10) = 0.27$ J

1–11 $i_2(t) = -0.033 \cos (5t)$, $p_1(t) = p_2(t) = 10 \sin (5t) \cos (5t)$

1–13 $p = 100R_L/(5.4 + R_L)$, plotted in Fig. S–3, R $= 5.4$ Ω

1–15 See Fig. S–4.

1–17 $i(t) = 2.5t$, $v(t) = 250t$, $p(t) = 625t^2$

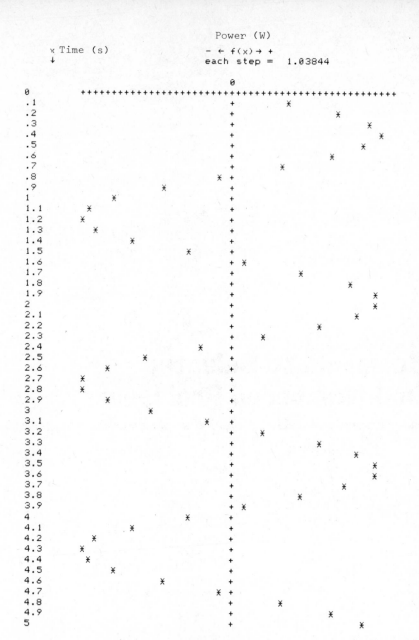

FIGURE S-1.

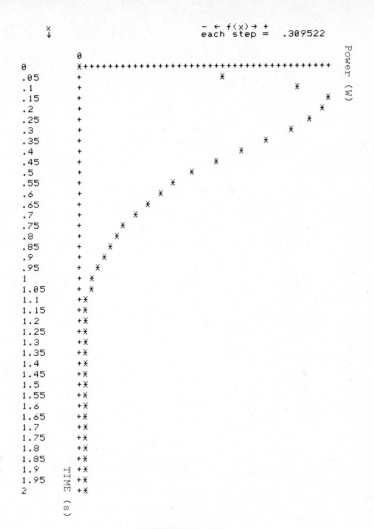

FIGURE S–2 (a).

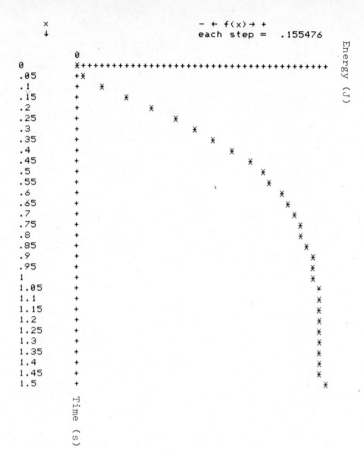

FIGURE S—2 (b).

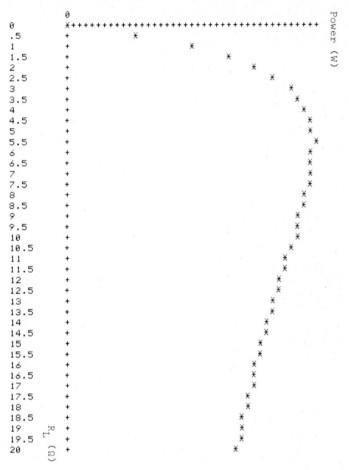

FIGURE S–3.

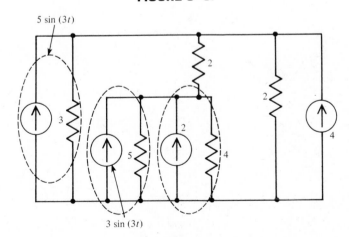

FIGURE S–4.

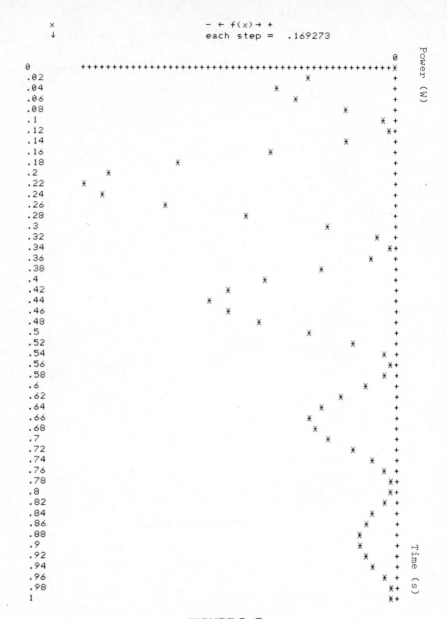

FIGURE S-5.

1–19 $V_1 = 3e^{-2t} - 7 - 6 \cos (4t)$
   $V_2 = 11 + 6 \cos (4t) - 3e^{-2t}$
   $V_3 = 2 + 6 \cos (4t)$

1–21 $v(t) = 4 + 4 \sin (\pi t/4)$
   $v(t) = 3 \cos (\pi t/2 - \pi/4)$ or $3 \sin (\pi t/2 + \pi/4)$
   $v(t) = 7 - 9e^{-1.69t}$

1-23 $i(1.5) = -0.278$ A, $p(1.5) = 1.55$ W
$i(4.5) = -0.088$ A, $p(4.5) = 0.156$ W

1-25   Plotted in Fig. S–5. Maximum is 8.49 W at 0.22 s by the plot. This could be refined by better plotting or calculus.

Chapter 2

2-1   For Fig. 2–34a: $i_{5\Omega} = 1.170$ A counterclockwise,

$$i_{3\Omega} = 0.383 \text{ A counterclockwise,}$$

$$i_{4\Omega} = 0.787 \text{ up}$$

For Fig. 2–34b: $i_{2\Omega} = 1/2 + e^{-2t}/4$ down (resistor on left)

$$i_{2\Omega} = 2/3 - e^{-2t}/3 \text{ down}$$

$$i_{6\Omega} = 1/6 - 7e^{-2t}/12 \text{ up}$$

$$i_{4\Omega} = 2 + 2/3 - e^{-2t}/3 \text{ up}$$

2-3 $a_{12} = -4$, $a_{21} = 2$, $a_{34} = 1$, $a_{23} = -4$

2-5   **(a)** $\begin{pmatrix} 2 & -1 \\ 12 & 2 \end{pmatrix}$ **(b)** $\begin{pmatrix} 13 & 10 \\ 0 & 4 \\ 1 & 4 \end{pmatrix}$ **(c)** $\begin{pmatrix} 4 & -5 \\ 11 & 16 \\ 21 & 42 \end{pmatrix}$

2-7 $\begin{pmatrix} 6 & -2 & -4 & 2 \\ 0 & 2 & 4 & -1 \\ 3 & 0 & -2 & 4 \\ -6 & -4 & 1 & -3 \end{pmatrix} \begin{pmatrix} x \\ y \\ z \\ w \end{pmatrix} = \begin{pmatrix} -4 \\ 0 \\ -7 \\ 14 \end{pmatrix}$

2-9   **(a)** 0; **(b)** -58; **(c)** $32 \cos(2t) + 8e^{-3t} \cos(2t) - 77$

2-11   DET $= -132$

2-13 $x = -0.489$, $y = -1.809$, $z = 0.638$, $w = -1.064$

2-15   **(a)** $i_i = 0.444 - 0.\cos(5t)$, $i_2 = -1.213 + 0.312 \cos(5t)$, $i_3 = 0.724 + 0.114 \cos(5t)$; **(b)** $i_1 = -0.096$, $i_2 = 0.154$

2-17 $\begin{pmatrix} 14 & -2 & 0 & -8 & 0 \\ -2 & 18 & -5 & 0 & -7 \\ 0 & -5 & 26 & 0 & -5 \\ -8 & 0 & 0 & 15 & -3 \\ 0 & -7 & -5 & -3 & 17 \end{pmatrix} \begin{pmatrix} i_1 \\ i_2 \\ i_3 \\ i_4 \\ i_5 \end{pmatrix} = \begin{pmatrix} 1 \\ 9 \\ 13 \\ -2 \\ -6 \end{pmatrix}$

2-19 $i_{2\Omega} = 2.011 \cos(5t) - 1.848$

FIGURE S–6.

2–21 $v_{out} = 13.158 + 4.737 \cos (4t)$, gain = 2.369

2–23 $i_N = 11.333 + 16.667 \cos (5t)$, $R_N = 2.065 \ \Omega$

2–25 $R_{TH} = 5.023 \ \Omega$, $v_{TH} = 0.497 \ V$ positive at point a

2–27 $R = 2.125 \ \Omega$

2–29 $i_1 = 0.17898$, $i_2 = 0.77280$, $i_3 = 0.68024$,
$i_4 = -0.00498$, $i_5 = 0.164463$

2–31 $v_{out} = 0.25 \sin (5t) + 4.305e^{-3t}$, plotted in Fig. S–6

Chapter 3

3–1 $7i + 2(di/dt) + (1/4) \int i \, dt = 2 \sin (5t) - 5$

3–3 $6(dv/dt) + (1/4) \int v \, dt + 6v/10 = 5$

3–5 $i_{4H}(0) = 0.55$ A, up; $v_{4F}(0) = -2$ V, + top

3–7 $i_{4H} = 0.759$ A, up; $v_{4F} = 6.107$ V, $-$ top

3–9 $\dfrac{v_1}{5} + 6\dfrac{dv_1}{dt} + \displaystyle\int v_1 \, dt - \int v_2 \, dt = 2$

$-v_1 \, dt + \dfrac{19}{12} \displaystyle\int v_2 \, dt - \dfrac{1}{4} \int v_3 \, dt = 0$

$-\dfrac{1}{4} v_2 \displaystyle\int dt + \dfrac{v_3}{2} + 4\dfrac{dv_3}{dt} + \dfrac{1}{4} \int v_3 \, dt = 0$

3–11 $9i_1 - 5i_2 = 2e^{-3t} - 2$

$5i_2 + 8\dfrac{di_2}{dt} + 4\dfrac{di_3}{dt} = 2$

$4\dfrac{di_1}{dt} + 10i_3 + 8\dfrac{di_3}{dt} + \dfrac{1}{4} \displaystyle\int i_3 \, dt = -6$

3–13 $v(t) = 5(1 - e^{-t})$

3–15 $i(t) = 0.676e^{-0.007t} - 0.676e^{-0.993t}$

3–17 $i(t) = -0.733e^{-0.036t} + 0.867e^{-3.46t} - 0.134 \cos (5t)$
$+ 0.095 \sin (5t)$; plotted in Fig. S–7

3–21   For $0 < t < 3$: $i(t) = 2.58e^{-t/4} \sin (0.968t)$
For $t > 3$:

$$i(t) = -1.39e^{-t/4} \sin (0.968t) + 0.287e^{-t/4} \cos (0.968t)$$

Plotted in Fig. S–8.

3–23 $i_1(t) = 1.49e^{-2.56t} + 0.065e^{-t/3} - 1.51 \cos (3t)$
$- 2.79 \sin (3t)$

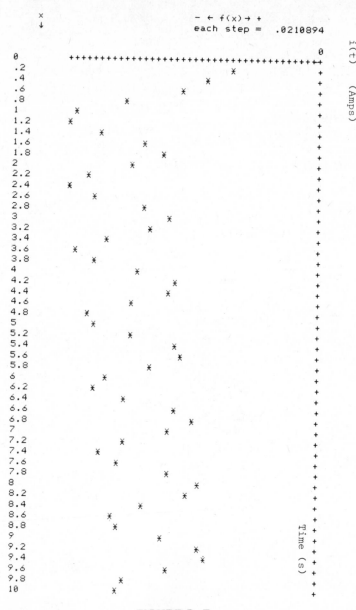

FIGURE S–7.

$$i_2(t) = -0.72e^{-2.56t} + 0.108e^{-t/3} + 0.612 \cos (3t)$$
$$- 0.588 \sin (3t)$$

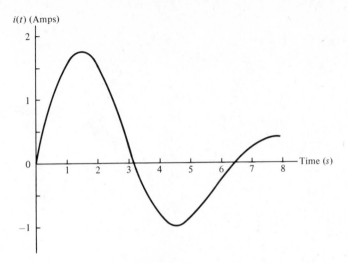

$i(t)$ (Amps)

Time (s)

FIGURE S-8.

Chapter 4

4-1 $F(s) = 5/s^2$

4-3   (a) $f(t) = 4\delta(t) + 4$; (b) $f(t) = -6 + 3e^{-7t}$;
(c) $f(t) = 3\cos(3t) + 9\sin(3t)$; (d) $f(t) = 6e^{-2t}\cos(4t)$;
(e) $f(t) = 0.4e^{-4t}\sin(5t)$

4-5 $F(s) = 3/s^2 - 24/(s^2 + 16)$ and $d/dt \rightarrow s$
$df/dt \rightarrow sF(s) = 3/s - 24s/(s^2 + 16)$
Inverting; $df/dt = 3 - 24\cos(4t)$

4-7   (a) $(10I + 3s)I_1 - (5 + 3s)I_2 = 2s/(s^2 + 9)$
   $- (5 + 3s)I_1 + (5 + 3s + 1/2s)I_2 - I_3/(2s) = 0$
   $-I_2/(2s) + (4 + 1/(2s))I_3 = 0$
(b) $(2 + 4s)I_1 - 2sI_2 = 4/(s^2 + 4)$
   $- 2sI_1 + (5 + 6s)I_2 = -10/s$
(c) $(0.1 + 4s + 0.125/s)V_1 - 0.125V_2/s = 20/(s^2 + 16)$
   $-0.125V_1/s + 0.025V_2/x - 0.1V_3/s = 0$
   $-0.1V_2/s + (6s + 0.1/s)V_3 = 0$

4-9   (a) dc level; (b) e^{-6t}; (c) te^{-6t}; (d) $\sin(2t)$;
(e) $\cos(2t)$; (f) e^{-t}

4-11   (a) $-0.56 + 0.9e^{-2t} - 0.34\cos(4t) + 0.45\sin(4t)$
(b) $0.21e^{-t} + 0.79e^{-4t}\cos(5t) - 0.72e^{-4t}\sin(5t)$
(c) $0.11 + 1.89\cos(3t) - 1.33\sin(3t)$
(d) $2\delta(t) - 0.2 - 3.8e^{-t}\cos(2t) - 2.9e^{-t}\sin(2t)$

```
        x                      -  ← f(x) →  +
        ↓                      each step =   .529126


           0
   0       ++++++++++++++++++++++++++++++++++++++++++++++++++++++++    v(t)
   2       +                                                    *
   4       +                                                    *
   6       +                                                  . *         (Volts)
   8       +                                                    *
  10       +                                                    *
  12       +                                                    *
  14       +                                                    *
  16       +                                                    *
  18       +                                                     *
  20       +                                                     *
  22       +                                                     *
  24       +                                                     *
  26       +                                                     *
  28       +                                                     *
  30       +                                                     *
  32       +                                                      *
  34       +                                                      *
  36       +                                                      *
  38       +                                                      *
  40       +                                                      *
  42       +                                                      *
  44       +                                                      *
  46       +                                                      *
  48       +                                                      *
  50       +                                                      *
  52       +                                                      *
  54       +                                                       *
  56       +                                                       *
  58       +                                                       *
  60       +                                                       *
  62       +                                                       *
  64       +                                                       *
  66       +                                                       *
  68       +                                                       *
  70       +                                                       *
  72       +                                                       *
  74       +                                                       *
  76       +                                                       *
  78       +                                                       *
  80       +                                                       *
  82       +                                                       *
  84       +                                                       *
  86       +                                                       *
  88     T +                                                       *
  90     i +                                                       *
  92     m +                                                       *
  94     e +                                                       *
  96       +                                                       *
  98    (s)+                                                       *
 100       +                                                        *
```

FIGURE S–9.

4–13 $f(t) = -0.0069 - 0.056e^{-t} + 0.064e^{-6t} + 0.189te^{6t}$
$+ 0.071 \sin (2t) - 0.0015 \cos (2t)$

4–15 $i(t) = -1.67e^{-2.5t} \cos (4.33t) + 0.92e^{-2.5t} \sin (4.33t)$, transient

4–17 $v(t) = 26.73 - 2.73e^{-0.23t}$, plotted in Fig. S–9

4–19 $(A^2 + B^2)^{1/2} = 2\omega/(\omega^4 + \omega^2/2 + 1/16)^{1/2}$, plotted in Fig. S–10

4–21   Zeros: 1.90, -2.86, $-0.02 + 2.1j$, $-0.02 - 2.1j$
      Poles: -4, $-4.93 \pm 6.07j$, $-2.91 \pm 0.94j$

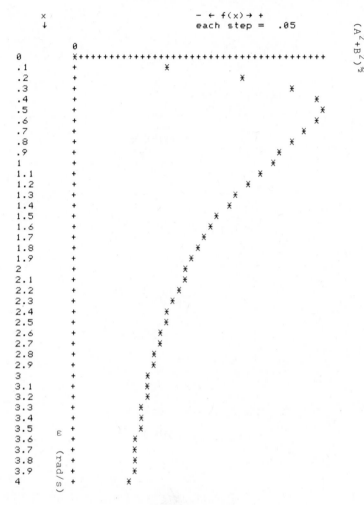

FIGURE S–10.

Chapter 5

5–3   See Fig. S–11.

5–5   See Fig. S–12.

5–7
$$\begin{pmatrix} 1.5 & -1 & 0 \\ -1 & 1 + 2/s & -2/s \\ 0 & -2/s & 0.5 + s/5 + 2/s \end{pmatrix} \begin{pmatrix} V_1 \\ V_2 \\ V_3 \end{pmatrix} = \begin{pmatrix} -2.5/s \\ 1/s \\ 0 \end{pmatrix}$$

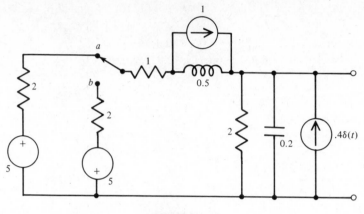

FIGURE S–11.

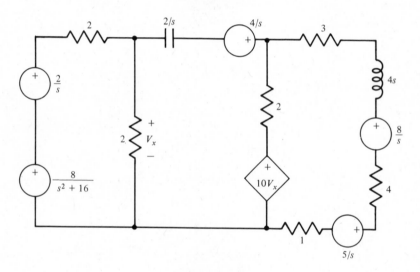

FIGURE S–12.

5–9 $v_2(t) = -2 + e^{-4.25t} [4 \cos (2.63t) + 6.45 \sin (2.63t)]$; plotted in Fig. S–13.

5–11 $v_{out}(t) = 0.01 \cos (2t) + 0.08 \sin (2t) + 0.03e^{-2.9t}$
$+ e^{-0/08t} [0.16 \sin (0.28t) - 0.05 \cos (0.28t)]$

5–13 For $v(t) = A \sin (\omega t) + B \cos (\omega t)$ the amplitude becomes

$$(A^2 + B^2)^{1/2} = 5(25 - \omega^2)[625\omega^2 + (25 - \omega^2)^2]^{1/2}/D$$
$$D = \omega^4 + 575\omega^2 + 625$$

Plotted in Fig S–14.

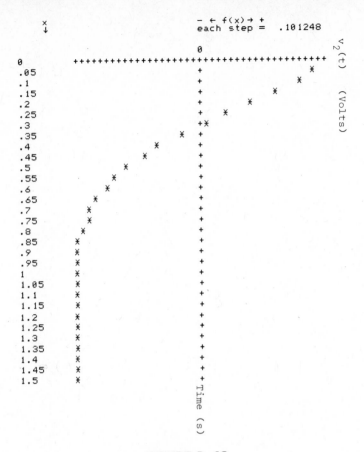

FIGURE S-13.

5-15 For $K < \sqrt{80}$: $v(t) = e^{-at}[A \sin (\omega t) + B \cos (\omega T)]$
For $K = \sqrt{80}$: $v(t) = K_1 e^{-at} + K_2 t e^{-at}$
For $K > \sqrt{80}$: $v(t) = K_1 e^{-a_1 t} + K_2 e^{-a_2 t}$
Examples:

$$K = \sqrt{80}: \quad v(t) = -1.41 + 10t + 1.41 e^{-114.14t} + 10t e^{-14.14t}$$

$$K = 2: \quad v(t) = -1 + 10t + e^{-10t} \cos (10t)$$

$$K = 3: \quad v(t) = -1.5 + 10t - 0.5 e^{-20t} + 2e^{-10t}$$

5-17 $v(t) = -0.255 e^{-0.063t} - 0.382 e^{-0.121t} \cos (0.123t)$
$+ 1.381 e^{-0.121} \sin (0.123t) + 5.005 \cos (5t) - 0.518 \sin (5t)$
Plotted in Fig. S-15.

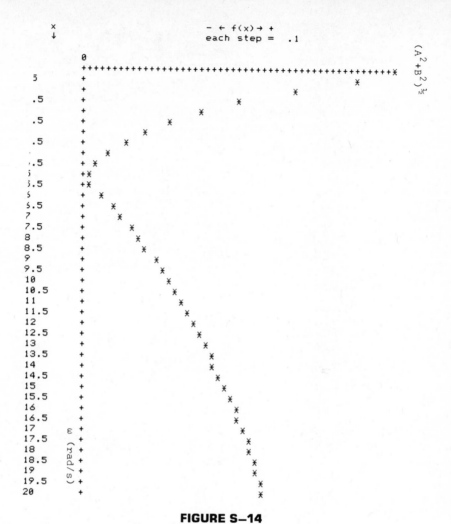

FIGURE S—14

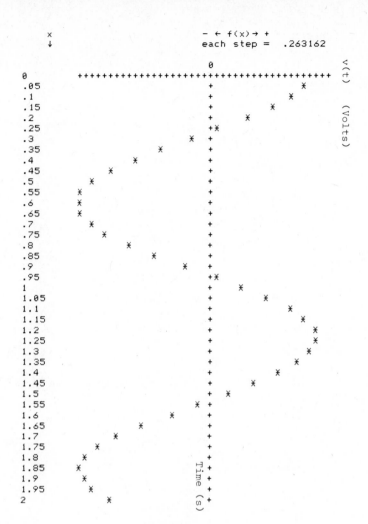

FIGURE S–15

Chapter 6

6–1 (a) Need six terms; (b) plotted in Fig. S–16

6–5 (a) $T = 3$; even, so $b_n = 0$
     (b) $T = 2$; odd, so $a_n = 0$ and $a_0 = 0$
     (c) $T = 4$; no time symmetry but a_0 and 0 by level symmetry
     (d) $T = 5$; even, so $b_n = 0$

6–7 (a) $f(t) = 4/3 + 1.82 \cos (2\pi t/3) + 0.46 \cos (4\pi t/3)$
         $+ 0.11 \cos (8\pi t/3)$
        Line spectrum plotted in Fig. S–17a.
     (b) $f(t) = 1.27 \sin (\pi t) + 0.64 \sin (2\pi t) + 0.42 \sin (3\pi t)$
        Line spectrum plotted in Fig. S–17b.

6–9 $f(t) = -2.12 \cos (\pi t) - 0.42 \cos (2\pi t) - 0.18 \cos (3\pi t)$
     Line spectrum plotted in Fig. S–18.

6–11 $v(t) = 3.18 - 2.12 \cos (\pi t) - 0.24 \cos (2\pi t) - 0.06 \cos (3\pi t)$
     Plotted in Fig. S–19.

6–13 Output plotted in Fig. S–20; phase and gain plotted in Fig. S–21.

6–15 Line spectrum plotted in Fig. S–22.

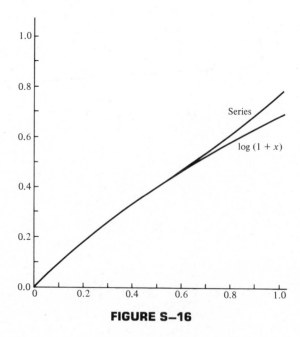

FIGURE S–16

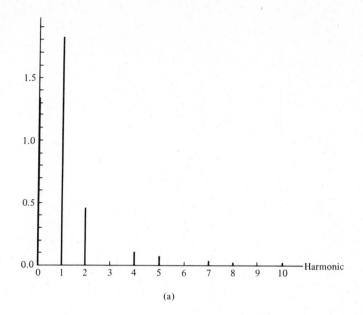

(a)

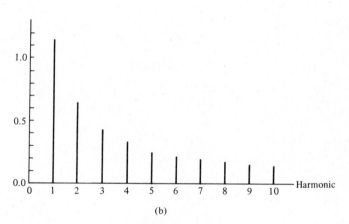

(b)

FIGURE S–17

6–17 Spectrum plotted in Fig. S–23.

6–19 Spectrum plotted in Fig. S–24.

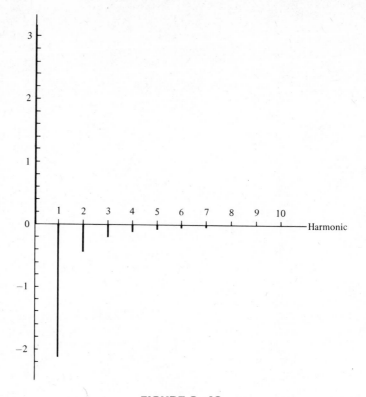

FIGURE S–18

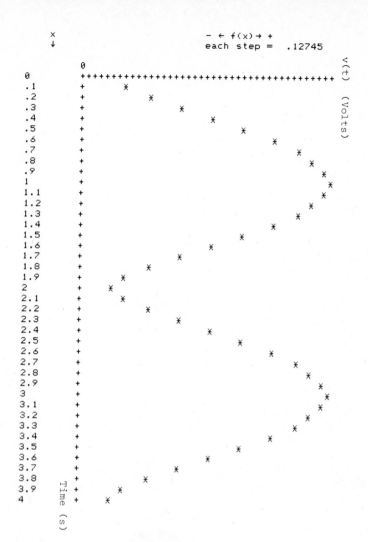

FIGURE S–19

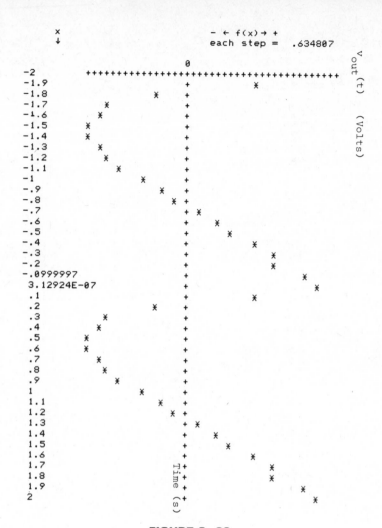

FIGURE S–20

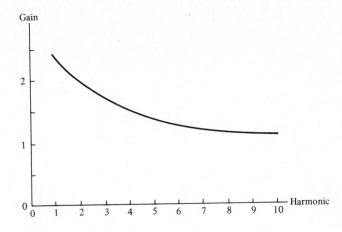

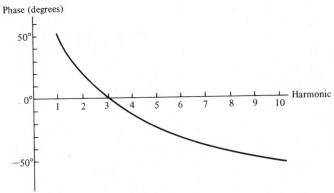

FIGURE S–21

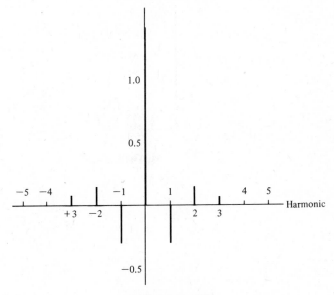

FIGURE S–22

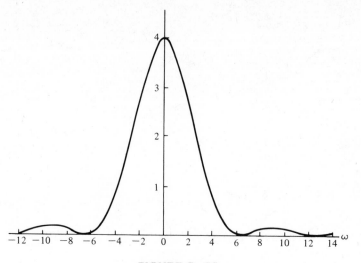

FIGURE S–23

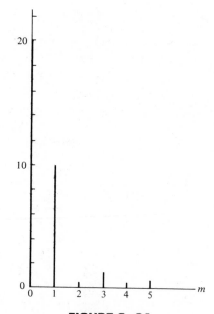

FIGURE S–24

Index